GETTING STARTED IN WOODWORKING™

Your First Workshop

A Practical Guide to What You Really Need

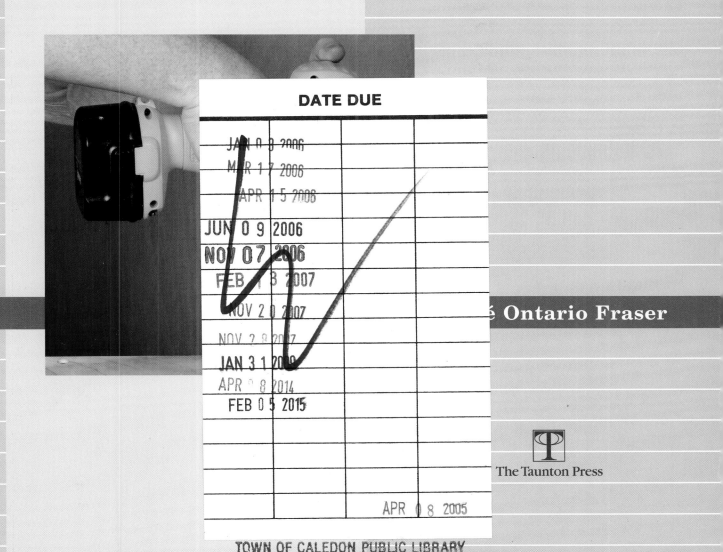

é Ontario Fraser

The Taunton Press

**To the makers and builders,
and especially those who want to be**

Text © 2005 by Aimé Ontario Fraser
Photographs © 2005 by Philip Dutton
Illustrations © 2005 by The Taunton Press, Inc.

The Taunton Press
Inspiration for hands-on living®

The Taunton Press, Inc., 63 South Main Street, PO Box 5506, Newtown, CT 06470-5506
e-mail: tp@taunton.com

Editor: Matthew Teague
Jacket/cover design: Michael Sund
Interior design and layout: Barbara Balch
Illustrator: Chuck Lockhart
Photographer: Philip Dutton

Getting Started in Woodworking™ is a trademark of The Taunton Press, Inc.,
registered in the U.S. Patent and Trademark Office.

Library of Congress Cataloging-in-Publication Data

Fraser, Aimé Ontario.
  Your first workshop : a practical guide to what you really need / Aimé Ontario Fraser.
    p. cm. -- (Getting started in woodworking)
  ISBN 1-56158-688-9
  1. Woodwork. I. Title. II. Series.
  TT180.F735 2005
  684'.08--dc22

                        2004019707

Printed in the United States of America
10 9 8 7 6 5 4 3 2 1

The following manufacturers/names appearing in *Your First Workshop* are trademarks:
Benadryl®, Betadine®, Delta® Machinery, GenTeal®, Kevlar®, Krazy Glue®, Masonite®, Nicholson®,
Quick-Grip®, Speed® Square, Spirograph®, Tapcon®, Vise-Grip®.

ABOUT YOUR SAFETY
Working with wood is inherently dangerous. Using hand or power tools improperly or ignoring safety
practices can lead to permanent injury or even death. Don't try to perform operations you learn about
here (or elsewhere) unless you're certain they are safe for you. If something about an operation doesn't
feel right, don't do it. Look for another way. We want you to enjoy the craft, so please keep safety fore-
most in your mind whenever you're in the shop.

Acknowledgments

Writing a book is a big task, but it is only a small part of the total work required to get one to publication. I owe many people debts of gratitude for their help along the way.

The greatest is to the team at The Taunton Press, who helped shape my vision of a shop book into something that works. My friend and sometime teaching partner Helen Albert worked closely with me developing what she liked to call "the stages of shopness."

Designer Rosalind Wanke often joined us and later spent a lot of time helping me think and write in two-page spreads. This book's visual impact is proof of her artistic genius. Carolyn Mandarano, as soothing as ever, kept me on track by pushing me onward in a way I didn't even notice at the time.

Editor Matthew Teague, a talented woodworker and writer himself, improved my words and was a pleasure to work with. Photographer Phil Dutton worked hard to capture the essence of the tools and their operations, and he made our long photo sessions fun with his gift of conversation.

Getting all those tools and moving them around, within, and between the two shops we used for photography was a lot of hard work I'm glad I didn't have to do alone.

Harry Brennan, my local Delta® Machinery representative, went out of his way more than once to make sure we had what we needed. His hard work was crucial to getting the book done on time. Angie Shelton of Delta worked with Harry to move a couple of tons of iron my way for photos. Wally Wilson of Lee Valley Tools helped me out more than once. Steve DeMonico, the hardest working man in woodworking, gave his usual product and moral support. John and Ginny Matchak, owners of The Woodworkers Club in Norwalk, Conn., ordered me tools, sold me tools, loaned me tools, and often called up just to see if I needed anything that day. Pat Carroll and Frank Colcone served as my muscle guys, helping move a shop full of tools across town at least three times. Martin Mittag, owner of The Wooden Boat Workshop of Norwalk, Conn., was patient and gracious when photography in his shop took longer than expected.

Finally, to Lois, who cheerfully endured my marathon photo and writing sessions and whose consistent care for me and enthusiasm for the project made it all possible.

—Aimé Ontario Fraser, Westport, Conn.

Contents

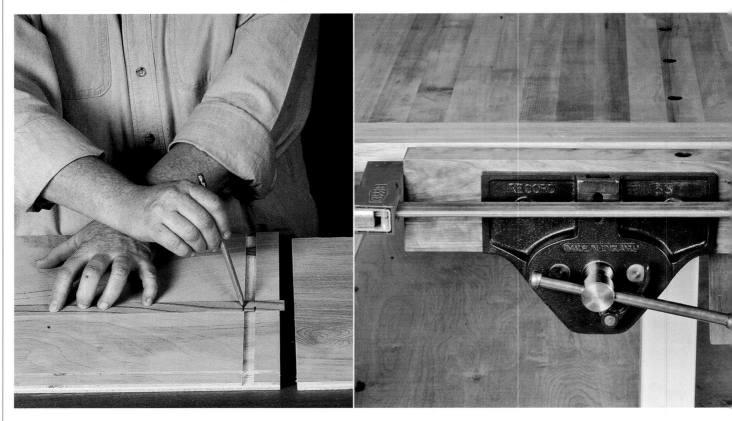

■ The Efficient Shop 90

■ The Well-Rounded Shop 130

Introduction

Woodworking matters. It's more than a pastime or hobby—being a woodworker means that you know the satisfaction and pride that comes from using your hands and mind to build beautiful, functional objects, and that you're as interested in the process as the outcome. Amid the speed and chaos of the modern world, woodworking gives us a place where we can slow down, pay attention, and take the time to do things right. Woodworking matters because it can make your life richer and more meaningful.

In woodworking, tools matter. It took a while for me to realize this because I had the not uncommon notion that if you had the right attitude, you could build a chest of drawers with rusty tools from the neighbor's shed. After some experience with decent tools, I realized that a properly sharpened and tuned plane is just as important as attitude. Good tools won't make you a great crafts-

person, but they will make it much easier to develop the skills needed to become one.

Your shop, the place where you keep and use your tools, is itself a kind of tool. A poorly laid out or unorganized shop can hinder the quality of your work just as surely as inferior tools. But your shop is more than a tool—it's also a creative studio where ideas become objects. For most of us, our shop is also a retreat where we can relax and recharge.

How this book is organized

This book recognizes that your skills as a woodworker, your collection of tools, and the layout and organization of your shop develop together. It's based on the notion that woodworkers go through four stages of development, and each stage has its own requirements for tools and space (see "The Four Stages of Woodworking" on the facing page).

The book is divided into four sections—one for each stage. Each section opens with an introduction explaining why it contains particular tools. Then it focuses on each tool in turn. Photographs across the bottom of the page show what the tool can do, and a photo illustration of the tool points out important features. The text explains aspects of using the tool and tells you what to buy. The section closes with a discussion of setting up and organizing a shop to use the tools properly.

One short book can't tell you all you need to know to master each tool. But it can tell you how to purchase a quality tool with the right features. It can't show the latest tricked-out models of tools, but by sticking to

simple classic tools, it can give you enough information to evaluate new features on your own.

The four stages of woodworking discussed here are not strict guidelines and can't take into account all the tools used in woodworking subspecialties (instrument making, boatbuilding, cabinetmaking, and the like). I don't expect you to buy the tools exactly in the order given, but the order has a logic. If you buy the tools and learn to use them in more or less the order given, you'll avoid the common mistake of buying too much too soon. Using the wrong tool or using the right tool improperly can be unpleasant enough to turn you away from woodworking. By following the book's progression, you'll create a solid foundation of woodworking skills you can build on with confidence.

THE FOUR STAGES OF WOODWORKING

Each section of this book sets out the tools and shop accommodations woodworkers need at each stage:

THE ESSENTIAL SHOP

A new woodworker needs to start slowly, mastering one thing at a time. Unsure of his future in woodworking, he sensibly doesn't want to spend a lot of money or time setting up a shop. He needs only the minimum tools and shop space essential for success.

THE BASIC SHOP

The novice woodworker is willing to spend some time and money on tools and setting up her shop but realizes that she has some skill building to do before she can get the results she wants. At this

level, a woodworker needs a set of tools that can accomplish all the basic operations used in making high level work, and a shop with enough room to do it.

THE EFFICIENT SHOP

The proficient woodworker wants to be in control of the creative process by milling his own lumber, and he wants to do so efficiently. He needs a shop full of machinery and the tools to maintain it.

THE WELL-ROUNDED SHOP

The experienced woodworker is ready to round out her shop by adding a few well-chosen tools and by spending some time organizing and refining the shop so it suits her style of woodworking.

The Essential Shop

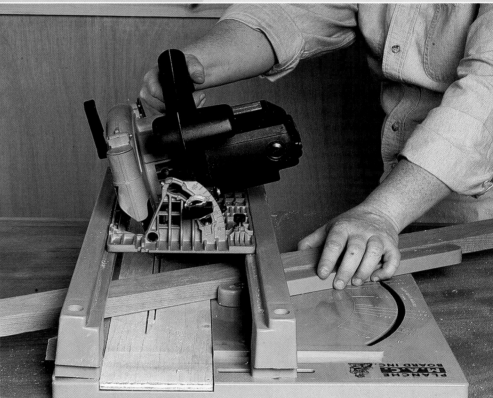

What to Consider

Though they share many operations, woodworking and carpentry are different trades. Carpenters work on site with a tool kit chosen for versatility and mobility. Woodworkers work at a bench in a shop (though they may share it with a car or hot-water heater) and use a larger collection of specialized tools capable of a wider range of operations and closer tolerances.

As you progress in woodworking, you'll find that there are some important differences between woodworking and carpentry tools. But while you're still learning basic skills, you can use the same tools a carpenter uses. They're readily available and less expensive, so you can get started in woodworking without a big outlay for tools. This section starts you out with the essentials—the tools you'll need to accomplish the most basic operations.

Laying a foundation

Your first tool should be the one that most clearly declares you a woodworker—a sturdy workbench. Determining the right bench is a delicate business, and many advanced woodworkers spend huge amounts of time obsessing over and building their perfect bench. New woodworkers need some experience before they can evaluate their bench needs, so the goal is to get something that meets the basic criteria without much fuss. Later, you can modify or retire it.

You'll also need a means of getting workpieces out of the lumber you buy, and a circular saw fits the bill. It's not elegant, but with the right setup, it does a good job with both solid wood and sheet goods. You'll need measuring tools for cutting out those parts and for getting them square, planes and chisels for fitting the pieces, and clamps to hold it all together while you fasten the joints using hammers, screwdrivers, and a cordless drill/driver. Once your projects are complete, sanding tools are used to smooth them before finishing. To keep you safe and efficient all the while, be sure to have a few cleaning tools and some proper protection devices.

One square, many uses. **Primarily a layout tool, a carpenter's speed square can also help you make certain your circular saw blade is set at 90°. If not, joints won't fit correctly.**

One step at a time

As you take on new projects and get familiar with your tools, you'll be learning new skills—how to stand when using the circular saw, how to drive a screw with a cordless drill/driver, and how to plane. Then you'll learn related skills—how to set up guides for the saw, how to sharpen your planes, and how to measure and mark efficiently and without error. As you do, you'll be learning how to think like a woodworker. You'll be learning the correct order in which to tackle operations, how to reduce mistakes, how to work efficiently, and more.

You're embarking on a craft that takes a lifetime to master. There's a lot going on, and it's important that you give yourself room for imperfection. Allot more time and materials than you think you'll need, and when something goes wrong, make the most of it. Before rushing off for the do-over, figure out just what went wrong and what you might have done to prevent it. You won't make that mistake again. With each one, you build up your store of woodworking wisdom. With each one, you get better.

With a good collection of clamps, you can hold the work to the workbench without the need for a vise. Eventually, you'll want a vise for the convenience of quickly securing and releasing parts.

In the beginning, before you've bought many tools, your shop space can be simple. You may even have room to share it with the car.

Workbench

The bench is the heart of any woodworking shop. From laying out and gluing up to planing, shaping, and finishing, it's where work gets done. The problem for the beginning woodworker is figuring out just what a good bench is. If you look to current woodworking literature for help, you'll see all kinds of benches. One author says you must have feature X, another suggests only Y. It can be overwhelming, especially when you don't have the experience to evaluate a design. That's why I suggest you set up your first shop with a simple bench you can modify to suit your work style as your skills increase.

What to buy

Every good woodworking bench, no matter what it looks like, meets a few basic criteria. First, it's sturdy enough to handle vigorous operations without moving. The legs are beefy, solidly joined, and well braced to anchor a thick, heavy top. If the bench lives in a shop

WATCH OUT

- Don't sand your benchtop—abrasives get embedded and scar your workpieces. Plane or scrape only.
- For the same reason, keep glue and finish off the bench by using a Masonite® cover.
- Applying finish to your benchtop makes it easier to keep clean.
- Don't round the bench edges or you'll lose the visual reference points.

with uneven floors, it needs large-diameter leveling glides. A convenient working height is between 33" and 36" tall, and the top should be between 20" and 32" wide—wide enough to hold workpieces, but not so wide you can't reach across it. Most range from 4' to 7' long, but length is not a critical issue.

Though it's possible to build light, stiff tabletops, you want weight in a bench. The

WHAT A BENCH CAN DO

■ BASE FOR CLAMPING

Assemble panels on the benchtop.

■ ASSEMBLY STATION

Use it as a steady base for applying edging.

■ FLAT, LEVEL WORK SURFACE

A bench provides a clear, stable surface for sanding.

traditional top is at least 2¼" thick, laminated from tough beech or maple. A similar thickness layered from manufactured materials like plywood, medium-density fiberboard (MDF), and core doors works nearly as well and is much easier to build. Just be sure that the top is flat so you're able to quickly clamp anywhere around the periphery and underside of the benchtop (see the photo on p. 10 as an example). That means no lengthwise stretchers or drawers near the top.

Build or buy?

Because your bench will be one of your first acquisitions when setting up shop, be conservative and choose a simple, sturdy bench that an inexperienced woodworker with minimal tools can build one.

Fashion a beefy base from well-chosen dry lumber or simply buy sturdy metal legs. Use a store-bought laminated slab for the top or build your own from plywood and MDF. The cost of building is moderate in time and money, and you can hardly go wrong. Even if you build another bench down the road, you can always use a sturdy horizontal

If you want to buy a simple bench, check with an industrial supply house. It won't look

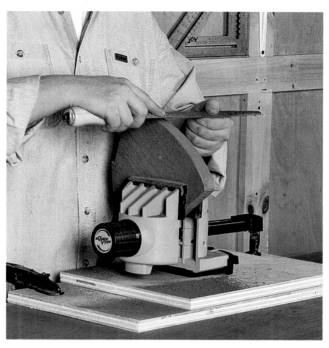

A portable vise **can be easily clamped in place to hold small complex workpieces at a convenient height for fine work.**

like a woodworking bench, but it'll do the job and cost less than building one. And though I can offer a strong argument for working your way up to a traditional cabinetmaker's bench, there's nothing to stop you from buying one from the start if you choose. Read ahead to the part on benches in the next section to learn more about your choices before buying.

■ ANCHORS JIGS

Use a hook to hold boards for crosscutting.

■ STABILIZE WORK

A corner can serve as a vise for sanding or light planing.

A Simple, Sturdy Bench to Start

Start out with a simple bench that doesn't cost much in time or money. As your woodworking skill develops, you can modify it to meet your needs. This bench was built in a few hours from a metal leg set and a solid-core exterior door.

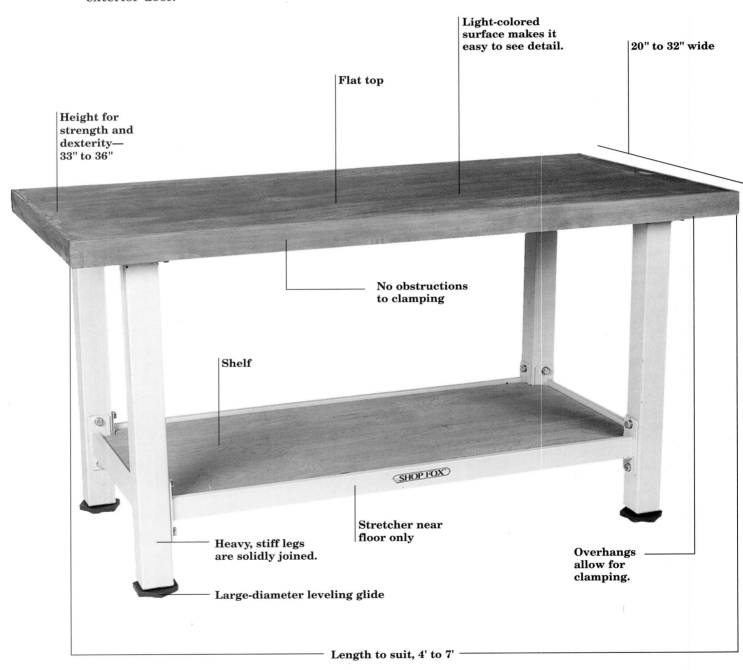

Light-colored surface makes it easy to see detail.

Flat top

20" to 32" wide

Height for strength and dexterity— 33" to 36"

No obstructions to clamping

Shelf

Heavy, stiff legs are solidly joined.

Stretcher near floor only

Overhangs allow for clamping.

Large-diameter leveling glide

Length to suit, 4' to 7'

SECURING WORK

CLAMP TO THE BENCHTOP
Whether you're sanding, drilling, or fastening, clamping a workpiece flat to the edge of the bench keeps it in place as you work. By using two clamps—one at each end—you'll prevent the workpiece from pivoting.

CLAMP HANDSCREWS IN PLACE
For working the edges of a board, it's often easiest to clamp the board upright in a handscrew, then clamp the handscrew to the bench. For long workpieces, use one handscrew at each end. Handscrews also do a good job holding odd-shaped pieces.

USE A BIRD'S-MOUTH VISE
Another method for working the edges of a board is to use a bird's-mouth vise. A simple V is cut into the edge of a board and then clamped to the bench. The workpiece simply wedges into the V. Work toward the V or clamp one at each end of the board.

CLAMP ACROSS THE FRONT
Wide boards can be clamped to the edges of the workbench with pipe clamps running underneath the benchtop. This approach is helpful for getting wide boards low enough so they're at a comfortable working height.

Cordless Drill/Driver

Cordless drill/drivers are light enough to use anywhere, have no power cord to tangle, and have the ability to drive screws without breaking them—no wonder corded drill sales are down. A wonderfully versatile tool, your cordless drill/driver will likely end up being the most-used power tool in your shop.

What to buy

Take the time to find a drill/driver that feels good in your hands. Balance, switch locations, and grip size are matters of taste, but they are more important to your satisfaction with the tool than motor ratings or foot-pounds of torque.

When selecting your drill/driver, don't be tempted to get the biggest one you can find. Stick with a 12- or 9-volt machine. They are lighter and easier to handle, and they have plenty of power to manage the jobs you do as a woodworker.

Two features that differentiate the cordless drill/driver from the mere drill are the high/low

speed switch and the adjustable clutch. As a rule, use the low speed for driving screws and boring large holes. High speed is for drilling small holes and removing screws efficiently.

The adjustable clutch disengages the drive when the torque reaches a certain level, so even though the trigger is fully depressed, the bit won't turn. You can set this feature to drive screws flush with the surface, deep below the surface, or anywhere between.

WHAT A DRILL/DRIVER CAN DO

■ BORE HOLES

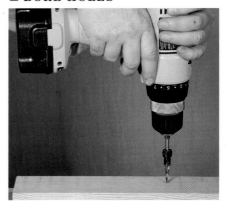

Use light pressure and stay in line with the hole.

■ DRILL WITH PRECISION

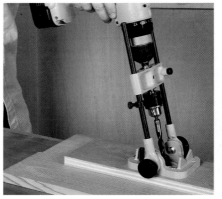

Jigs allow you to drill a hole at an exact angle—or location.

■ SAND WITH ATTACHMENTS

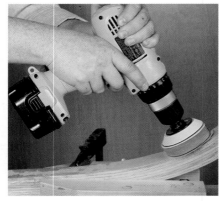

Disks and sleeves turn a drill into a sander.

Cordless Drill/Driver and Drill Bits

You'll use your cordless drill/driver more than any other power tool in the shop. Make sure your first has the features to satisfy your woodworking needs for years to come.

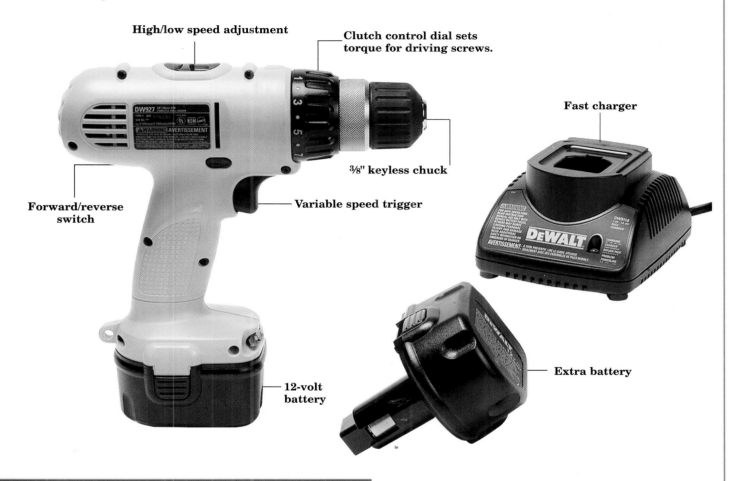

High/low speed adjustment

Clutch control dial sets torque for driving screws.

Fast charger

⅜" keyless chuck

Variable speed trigger

Forward/reverse switch

Extra battery

12-volt battery

■ CONSTRUCT STRONG JOINTS

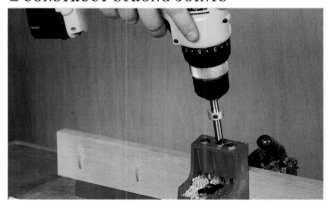

Drill and drive pocket hole screws easily.

You'll also want to get a drill/driver that has a reverse switch located conveniently near the trigger. You'll need two batteries to handle big jobs and a charger that can do its job in an hour or less so you're not stranded in the middle of a job. For best battery life, charge your batteries when you note reduced performance but before they're fully dead.

Set the speed to high, set the clutch in drill position, and fully depress the trigger.

Drill Bits

Whether you're drilling holes, driving screws, or using a specialized jig, there's a bit or drill to handle the job.

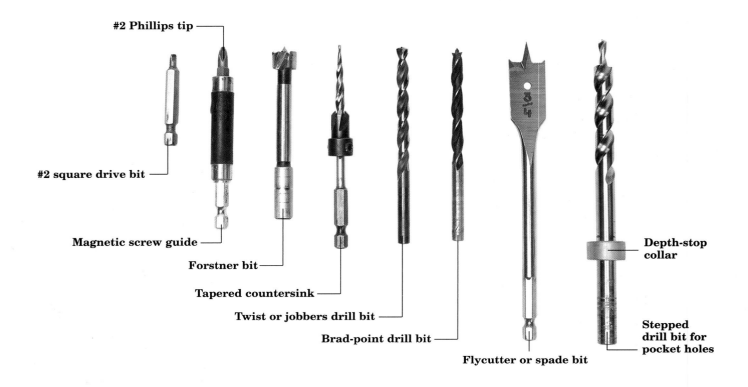

- #2 Phillips tip
- #2 square drive bit
- Magnetic screw guide
- Forstner bit
- Tapered countersink
- Twist or jobbers drill bit
- Brad-point drill bit
- Flycutter or spade bit
- Depth-stop collar
- Stepped drill bit for pocket holes

That's the gist of boring holes, but boring good, clean holes requires that you attend to a few other factors.

Don't lean on the drill while boring a hole—a sharp bit requires only a firm, steady push. With a little experience, you'll be able to hear a change in pitch when the drill is close to cutting through. At that point, lighten up and let the drill do the work. If you push too hard through the last bit of wood, you'll tear an ugly splinter from the back. For further insurance against tearout, hold or clamp the workpiece to a scrapwood backer board.

Choosing the right drill bit is crucial. A standard twist drill has a tendency to wander as it starts to cut, and the hole often winds up in the wrong place. You can reduce this tendency by punching a small dent or hole to direct the tip, but a better solution is to use brad-point bits designed for drilling wood. The sharp, protruding tip keeps the bit on course, and the bit's cutting angle helps sever wood fibers for clean, accurate cuts.

Carpenters, electricians, and other tradespeople commonly use a spade bit or fly-cutter in studs and joists. These inexpensive, disposable bits have their place, but Forstner bits are a better choice for fine work. They leave a clean, smooth surface because the outer rim scores the surface while the inner part is sliced clean. They're your best choice for smooth, large, or angled holes and for holes drilled to partial depth in a board.

To get started, buy a set of moderately priced twist drills in ¹⁄₆₄" increments up to ½". Beware of cheap sets, because the drills are often a little smaller than the measurement marked on their shanks. You'll also want brad-point bits—look for a set of at least seven. You can get by with an inexpensive set, but the more costly versions have a better tip design and cut more cleanly. For drilling pilot holes for screws, get at least five tapered countersink drills for the most common screw sizes. Finally, you'll want a set of Forstner bits in the common small sizes.

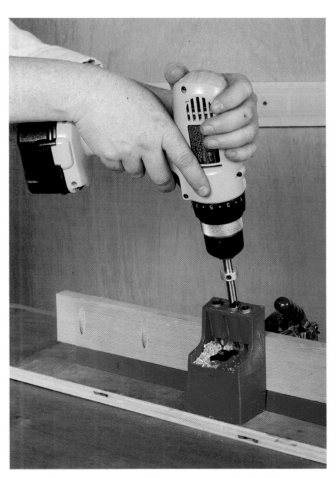

One joint, many uses. **Using a pocket-hole jig, you can join at right angles, edge join, and even fasten good-looking bevels.**

A hidden fastener. **A pocket-hole screwed joint is strong because the angled screw bites into long grain instead of weak end grain. The stepped drill used with pocket-hole jigs cuts a pilot hole for the screw shank and a shoulder to seat and conceal the screw head.**

Driving screws

The key to successful screwdriving is to remember that you must keep the bit aligned with the centerline of the screws and fully engaged. If you aren't positioned behind the drill and pushing hard, torque forces the bit upward and tears up the screw head.

Ideally, the screw pushes the wood out of its way as it goes in, but in dense hardwoods or near the ends of a board the wood can split. Prevent this by drilling a pilot hole slightly narrower than the screw's thinnest part.

Measuring and Marking Tools

There's a saying that carpenters measure to the nearest eighth of an inch, cabinetmakers to the nearest sixteenth, and boatbuilders to the nearest boat. Fussing with a ruler against the curves of a boat is hopeless—success comes from transferring lengths directly from the boat using wooden strips ticked with pencil marks at the correct locations. This cuts out common measuring errors and practically guarantees that workpieces laid out from the same tick strip are identical. It's an elegant solution—accurate and precise without taxing the brain. And it works on more than boats.

Measuring usually troubles new woodworkers one way or another. First, there's the problem of reading the scales. Then, it's hard

to get back in the groove of working with fractions. Many try to solve the problem by obsessing over measurements. A better solution is to avoid measuring whenever possible by using tick strips (see the photo below).

That's not to say you throw out your tape measure and rulers. You need them, but remember that it's all relative. You'll get your best results when you attend not to the perfection of each part but to the sum of the parts.

What to buy

The first tool you'll need is a compact tape measure, no longer than about 16'. Pick one with a simple scale that works for you—ones with big numbers for reading without glasses, left-handed tapes, or ones with the fractions clearly labeled.

Tapes are not good for measuring short distances, so you'll also need a 12" metal cabinetmaker's ruler. Get one that reads right to

Better than a ruler. A tick strip stores measurements without having to read a tape or decipher fractions. Mark the strip with the location of the groove. Move the tick strip to the other board, align the reference mark, and transfer the location.

Measure Twice, Cut Once

Without reliable measurements, project parts won't fit together. A few simple tools are all that is required.

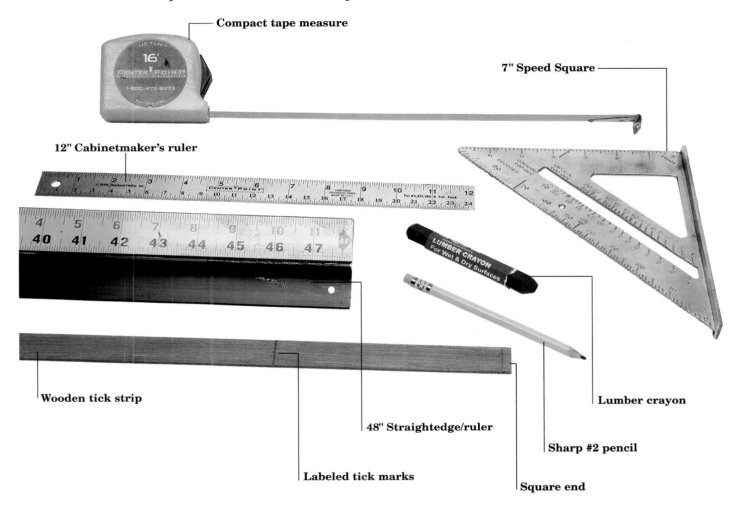

Compact tape measure

7" Speed Square

12" Cabinetmaker's ruler

Wooden tick strip

Lumber crayon

48" Straightedge/ruler

Sharp #2 pencil

Labeled tick marks

Square end

left on one side and left to right on the other, preferably with 8ths and 16ths on one edge, 32nds and 64ths on the other.

A stiff, inexpensive metal ruler 36" or 48" in length can double as a straightedge for checking the flatness of glued panels and planed surfaces. You'll also use a 7" Speed® Square for laying out perpendicular and 45° lines, squaring corners, measuring, and more.

Poor marking habits can make good measurements bad. A sharp #2 pencil makes a mark that's bold but not too thick. Learn to sharpen it with a pocket knife or keep a small sharpener in your pocket. For rough layouts to determine the best use of your lumber, use a lumber crayon in a color that contrasts with the wood.

Edge Tools

Planing is one of my favorite things about working with wood. There's nothing as pleasant as the gentle, rhythmic exercise of guiding a plane over a board, the tearing-silk sound of a sharp blade on wood, the shimmering grain revealed, and the fragrant shavings spiraling to the floor. It's a defining act of woodworking, and knowing how to do it well is a fundamental skill.

A well-tuned plane can remove shavings as thin as four ten-thousandths of an inch. With such control, it's easy to get joints that fit perfectly, something you'd be hard pressed to do by setting up a powerful machine by trial and error. A plane can flatten a panel, profile an edge, and smooth a rough board until it feels like polished glass.

You'll also need to know how to use a chisel, no matter how many machines you end up with. Use it for chopping out waste wood in lap joints or dovetails, or to pare tissue-paper-thin shavings to fit a joint. When you know

how to wield a chisel, you'll have the ability to join or shape wood any way you want.

Learning to use planes and chisels is a lot like learning to putt or play the piano. It's a physical skill that takes practice to master. No new golfer expects to be an expert putter without putting in some practice time. It's the same with mastering planes and chisels. Spend some time practicing.

WHAT EDGE TOOLS CAN DO

■ FLATTEN AND SMOOTH

Remove saw marks and level edges after gluing.

■ PROFILE EDGES

Rounding or chamfering can be done by hand.

Block Plane and Butt Chisels

A small, one-handed block plane should be your first edge-tool purchase, and then a set of carpenter's butt chisels. Add a jig to hold your blades steady when honing, and you're ready to get to work (see the sidebar on p. 21).

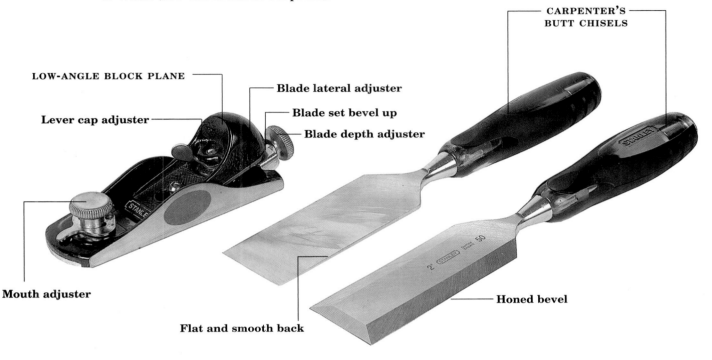

CARPENTER'S BUTT CHISELS

LOW-ANGLE BLOCK PLANE

Blade lateral adjuster

Lever cap adjuster

Blade set bevel up

Blade depth adjuster

Mouth adjuster

Honed bevel

Flat and smooth back

■ **PARE FINE SHAVINGS**

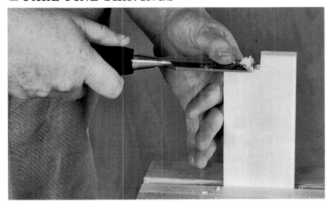

Remove tissue-thin slices for well-fitted joints.

■ **CHOP**

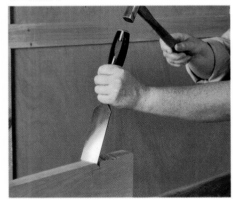

Remove waste with chisel and mallet.

Sharp tools make you a better woodworker. Sharpen after about 15 minutes of continuous use, or when it's no longer fun.

Before you can get anywhere with your edge-tool practice, you'll need to know how to sharpen. You'll never learn proper tool use or form with a dull tool—it fights you at every turn. Keep your tools sharp, and you'll be surprised at how quickly you become good at woodworking.

How often should you sharpen your edge tools? It depends on the job at hand, but the best indicator is your grumpiness level. When the tool is sharp, the work is satisfying and fun. A dull tool requires more effort, and it's not as easy to control. You'll start making mistakes, and wood that planed easily will begin to tear out. When you start thinking this isn't fun anymore, it's time to sharpen (usually after about 15 minutes of continuous work). Better yet, touch up the blade before you get grumpy.

What to buy

Your first plane should be a low-angle block plane with an adjustable mouth, readily available at hardware stores and home centers. These are the easiest to set up and maintain, and they are more versatile than their simpler cousins. You can open the mouth, or gap between the blade and the sole of the plane, to accommodate thick chips for rough work or close it down for fine shavings. Be sure to get a plane with a lateral adjuster so you can keep the blade parallel to the sole.

Start with a few carpenter's butt chisels. They're shorter and beefier than cabinetmaker's chisels and designed for hitting with a hammer. It makes sense to buy your first chisels as a set. Make sure that whatever you buy has at least ⅜", ½", and ¾" chisels. Round out your set with the widest chisel you can get—1½" or even 2".

You'll need a honing jig for holding the blade steady at the correct angle when sharpening. Which jig you get at this point is not as crucial as simply choosing one and using it as directed (see "Sharpening" on the facing page).

The best sharpening abrasive for beginning woodworkers is fine-grit wet or dry sandpaper glued to a piece of glass. Buy the sandpaper at an automotive-paint supply house in a range of grits—220, 320, 600, 1,000, 1,500, and 2,000. You can find spray adhesive at art supply stores.

Sharp tools make woodworking almost effortless, and sharpening is a task you must learn. The first step is to flatten the back of the plane or chisel blade. Lay the blade back-side down on the coarsest sandpaper in your sharpening kit (see the photo at top right), and move it back and forth until the whole surface is uniformly dull. Don't be surprised if this takes more than half an hour. Progress through your sandpapers from coarse to fine, spending three or four minutes on each until the blade's surface is shiny.

Then put the blade in your sharpening jig set to 30° and focus on the bevel, running it back and forth over your second-finest grit for about a minute (see the photo at center right). Once you can feel a little burr across the full width on the back side of the blade, go to your finest grit and hone for a minute or so. Finally, remove the burr by working the back a few times over the finest grit of sandpaper.

Check the edge by holding the blade lightly in one hand and letting it touch the thumbnail on the other hand. A sharp edge will catch on the nail. A dull one will skate over the surface.

Sharpening jig

A good jig gives a sharp edge every time.

You need only hone the very tip of the bevel.

Hammers and Screwdrivers

There's a tongue-in-cheek saying among carpenters and do-it-yourselfers that if the parts don't fit, get a bigger hammer. This is true in many situations, but not in woodworking, where finesse is the name of the game and the hammers are light. Instead of pounding and pulling nails, woodworkers use hammers for setting small nails, tapping fine joinery into alignment, and driving a chisel to remove excess wood.

Woodworkers use different screwdrivers, too. The familiar mechanic's screwdriver blade widens above the tip, while the cabinetmaker's driver is the same width from the tip well up the handle. Also, the tip on a woodworker's screwdriver is ground for a tight fit to prevent damage.

WATCH OUT

- Striking something with a hammer can make small pieces fly. Protect your eyes.
- Don't use a hammer with a loose head. Soak it until the wood swells and the head is tight on the handle again.
- Don't use cabinetmaker's screwdrivers around electricity—they won't protect you from a shock.
- While you may need to tap a screwdriver to loosen a screw, screwdrivers are not designed for other operations that require striking.

What to buy

Start your hammer collection with a light claw hammer meant for finish and trim work. Your second choice should be a small, light Warrington hammer with a cross peen for working inside a case or starting brads (see the photo at left). When you need more force, reach for a dead-blow hammer. Its head is half-filled with shot and oil so that the hammer doesn't bounce after the blow, and its soft coating prevents denting. For chiseling out waste, a well-balanced Japanese hammer weighing about 14 oz. works well. The slightly rounded face works for driving nails, and the flat face is for striking chisels.

Setting brads. Using the cross peen on a Warrington hammer, you can set even tiny brads without hurting your fingers. It's the ideal bench hammer.

Hammers for Woodworkers

Finesse is the key to woodworking, and heavy carpenter's hammers aren't necessary. Start your collection with a few light hammers, each suited to a specific purpose.

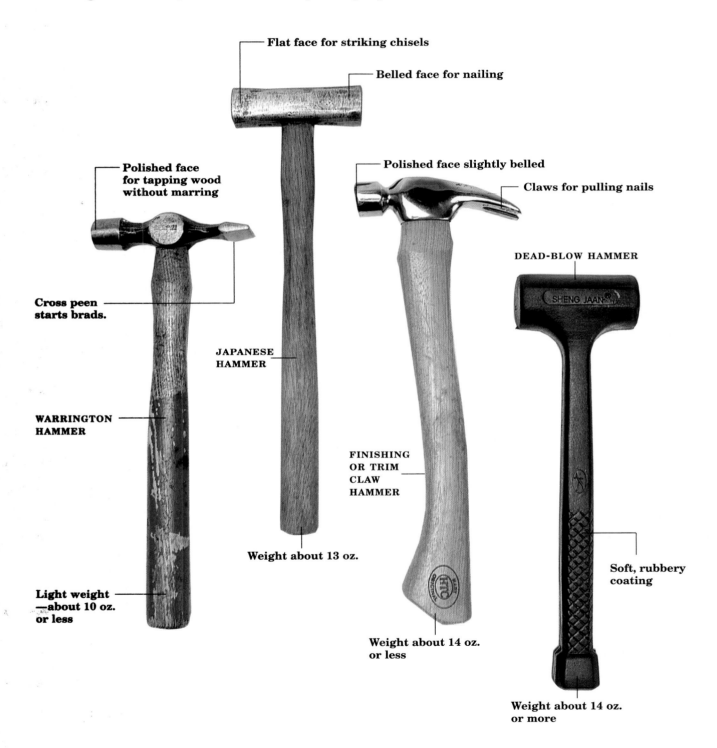

Flat face for striking chisels

Belled face for nailing

Polished face for tapping wood without marring

Polished face slightly belled

Claws for pulling nails

DEAD-BLOW HAMMER

Cross peen starts brads.

JAPANESE HAMMER

WARRINGTON HAMMER

FINISHING OR TRIM CLAW HAMMER

Weight about 13 oz.

Soft, rubbery coating

Light weight —about 10 oz. or less

Weight about 14 oz. or less

Weight about 14 oz. or more

Screwdrivers for Woodworkers

Y ou'll need mechanic's, cabinetmaker's, and Phillips screwdrivers in a range of sizes. Also keep on hand a good ratcheting driver with a wide selection of bits.

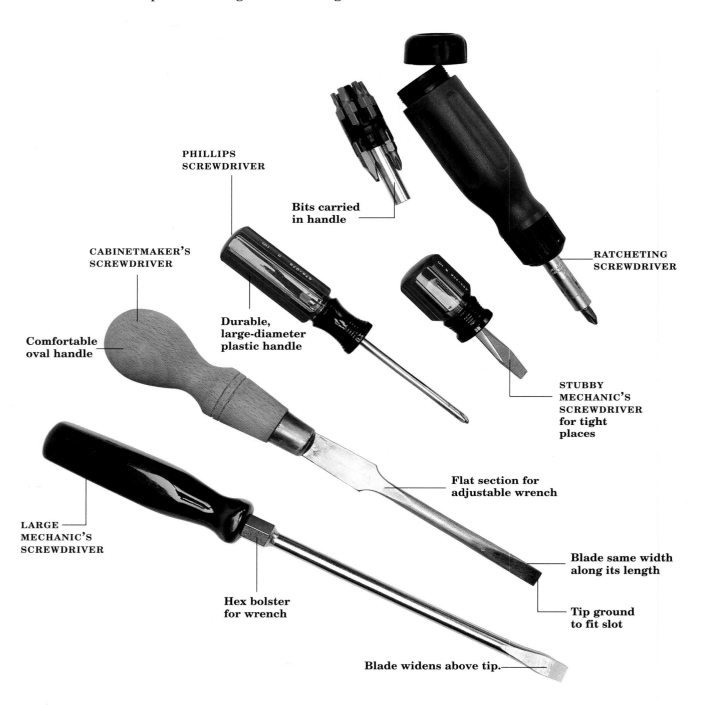

PHILLIPS
SCREWDRIVER

Bits carried
in handle

RATCHETING
SCREWDRIVER

CABINETMAKER'S
SCREWDRIVER

Durable,
large-diameter
plastic handle

Comfortable
oval handle

STUBBY
MECHANIC'S
SCREWDRIVER
for tight
places

Flat section for
adjustable wrench

LARGE
MECHANIC'S
SCREWDRIVER

Blade same width
along its length

Hex bolster
for wrench

Tip ground
to fit slot

Blade widens above tip.

Choose the right size screwdriver. The screw on the left was driven with a tip larger than the slot, leaving wood torn and ragged. The slot on the far right was chewed up by a blade that was too narrow. The screw in the middle was driven with the proper tip.

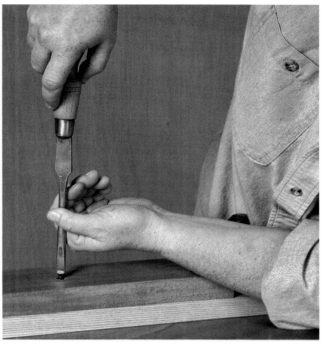

Approach is key. Driving a screw without damaging the head requires downward and rotating forces. Lean over and get your body into it. You can push harder and have more control if your elbow is behind the screwdriver.

You have two choices: A mechanic's screwdriver or a cabinetmaker's. The mechanic's screwdriver is the most familiar since it's cheaper to manufacture. The blade widens above the tip, and then tapers back to the shank diameter. Mechanic's drivers work well when the screw is set flush with the surface, but it's common practice in fine woodworking to drive screws below the surface and hide the head with a neat wooden plug. Because it's wider at the tip, the mechanic's screwdriver tears up the plug hole, ruining the tight fit and clean look. A cabinetmaker's screwdriver is the same diameter from tip to well up the shaft and never touches the surrounding wood. In addition, the tips of cabinetmaker's screwdrivers are carefully ground to fit the slot exactly.

Buy a set of proper cabinetmaker's screwdrivers. At the minimum, you'll need a #4, #6, #8, and #10. It's also worth getting a set of mechanic's drivers for rough work, including a stubby one for tight situations. You'll also need Phillips screwdrivers in the #2 and #3 size; #1 is less common but welcome if part of the set.

Assembling parts. A dead-blow mallet combines weight and a soft face. It's the best hammer for coaxing wooden parts into place.

Circular Saw

Properly handled, a circular saw can do the straight-line cutting jobs associated with chop saws and tablesaws, and with the right jigs and setups, you can do these jobs nearly as accurately. A good circular saw will take you a long way in woodworking, and even when you step up to stationary tools, you'll still use your circular saw for jobs like preparing rough boards for milling and cutting sheet goods.

What to buy

Your first circular saw should be a sidewinder with a 7¼"-diameter blade. It will cut boards up to 2¼" thick, but at about 10 lbs. it's not too heavy for most people. It's the most widely used circular saw, so you'll find a variety of blades and accessories readily available.

Until your tool collection grows to stationary power tools, you'll be asking a great deal of this saw. You'll want a corded saw's full power for cutting hardwoods, not to mention avoiding the hassle of running out of juice in the middle of a cut. Later, you can get a smaller battery-powered saw for convenience in light-duty applications.

> **WATCH OUT**
>
> - The best way to support a long or awkward workpiece for cutting is on a piece of foam insulation board.
> - Don't allow freehand cuts to wobble or the saw will kick back.
> - The saw tears out on the top of the cut, so put "good" side down.
> - Make sure the blade guard swings back in place before putting the saw down.
> - Set blade depth only ⅛" to ¼" deeper than the thickness of the wood to reduce the chance of kickback.
> - Don't pull backward on a running saw.

WHAT A CIRCULAR SAW CAN DO

■ CROSSCUT

Cut across the width of a board freehand.

■ RIP

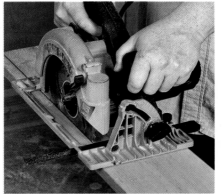

Cut along the length of a board.

■ BEVEL

Pivot the saw for angled cuts.

7¼" Sidewinder Circular Saw

This size and type of saw is the mostly commonly used, so accessories and blades are easy to find.

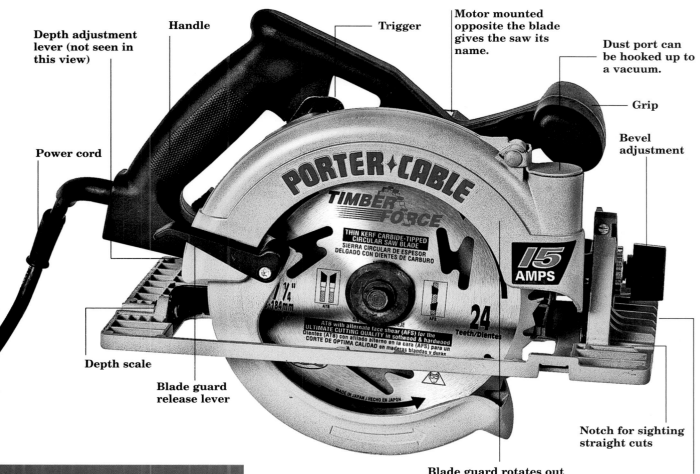

Depth adjustment lever (not seen in this view)

Handle

Trigger

Motor mounted opposite the blade gives the saw its name.

Dust port can be hooked up to a vacuum.

Grip

Bevel adjustment

Power cord

Depth scale

Blade guard release lever

Notch for sighting straight cuts

Blade guard rotates out of the way during the cut.

Stiff, sturdy sole

■ **PLUNGE**

Lower the blade into the center of a board.

The sole needs to be strong, straight, and flat. When comparing soles, think about how each one will stand up to being dropped or banged around.

Since you'll adjust the depth of cut frequently, choose a lock/unlock mechanism that is easy to operate, and look for a scale that is clear and readable.

Blade-right is the old standard, but blade-left models provide better visibility for right-handers.

Blades for Cutting Wood

To get the best performance from your circular saw, get a quality blade suited to the job at hand.

RIPPING
(16 to 20 teeth).
Fewer teeth means
quick clearing of
chips and less heat
buildup.

PLYWOOD
(90+ teeth). Many
teeth reduce chip-
out of veneered
surfaces.

CROSSCUT
(36 to 40 teeth).
More teeth leave a
smoother finish on
crosscuts.

COMBINATION
(18 to 24 teeth).
All-purpose use
includes ripping
and crosscutting.

FRAMING
(18 to 24 teeth).
Fewer, but more
specially shaped
teeth quickly cut
construction lumber.

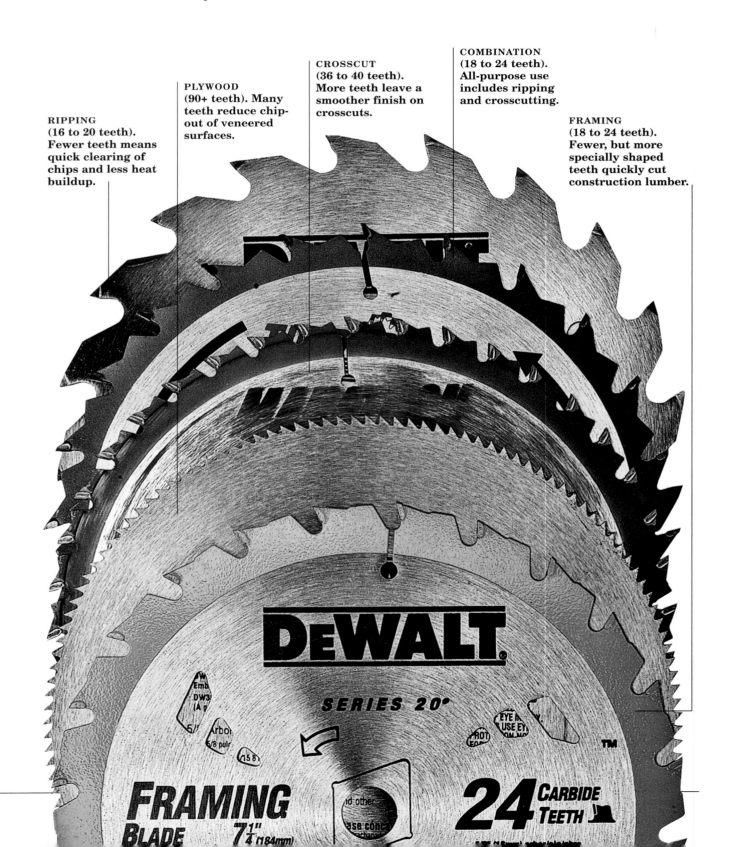

DEWALT

SERIES 20°

FRAMING
BLADE 7 1/4" (184mm)

24 CARBIDE TEETH

SQUARE OR PROTRACTOR

A large square with a lip held firmly against the edge of the workpiece can guide short crosscuts in solid wood or sheet goods.

RIP FENCE

Ideal for ripping from fairly narrow boards, a rip fence or edge guide uses the edge of the board to make a parallel cut. Set it on either side of the sole and run the saw on the wider piece for stability.

STRAIGHTEDGE FENCE

Commercial straightedges can be used for cutting sheet goods, ripping, wide cross-cuts, and plunge cuts—just about any type of straight cut. Get a commercially-made aluminum fence like this one, or build your own from plywood and Masonite.

CIRCULAR SAW MITER JIG

While it excels at cutting 45° miters, this miter box can also guide 90° crosscuts and compound angles. The saw runs on a track while you hold the workpiece at the desired angle.

Clamps

Here's a tip that will profoundly affect your woodworking life: Learn to use clamps. Most new woodworkers think clamps are just for holding stuff together while glue dries, but clamps are much more than that. They should be your main method for stabilizing your work. To prevent misaligned joinery, boards shifting as you work on them, and miscellaneous damage from dropping or bumping your work, use clamps.

What to buy

You'll need a varied clamp collection to meet common shop clamping situations. C-clamps are simple, strong, and inexpensive, and they work well in tight spots. Longer bar clamps are fast, versatile, and good for jobs requiring many clamps. Pipe clamps are indispensable when you need clamping power over long lengths. If your budget allows, you can supplement them with aluminum panel clamps for easy use in light-duty situations. For quick work that doesn't require extreme strength, Quick-Grip® clamps can serve as a convenient extra hand. And though handscrews require two hands to tighten, they are great for holding work that isn't parallel and are just about the only clamps you can clamp down to the bench.

WHAT CLAMPS CAN DO

■ JOIN BOARDS

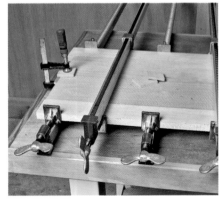

Panel clamps help keep glue-ups flat.

■ PROVIDE A THIRD HAND

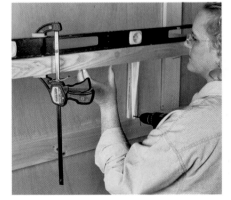

Clamps hold parts and tools steady.

■ GLUE UP

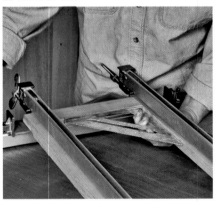

Square up a frame with proper clamping.

Basic Clamps

Start your clamp collection with a few well-chosen types that will serve most purposes and plan to buy more as needed for specific jobs. Having a few of each type is a good start.

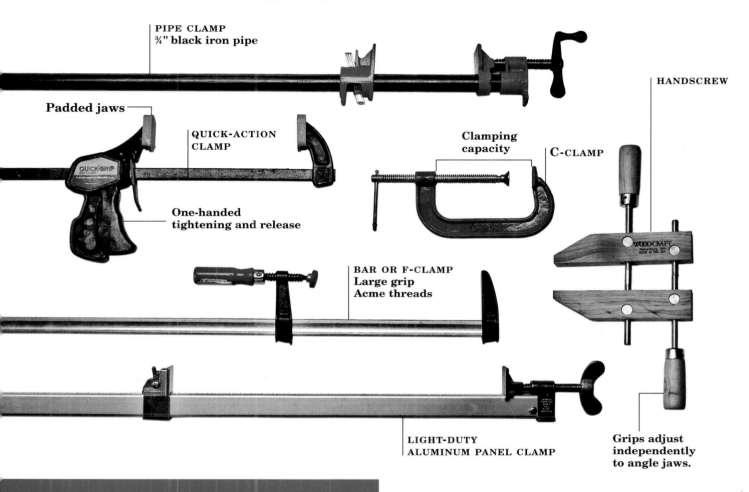

PIPE CLAMP
¾" black iron pipe

HANDSCREW

Padded jaws

QUICK-ACTION CLAMP

Clamping capacity

C-CLAMP

One-handed tightening and release

BAR OR F-CLAMP
Large grip
Acme threads

LIGHT-DUTY ALUMINUM PANEL CLAMP

Grips adjust independently to angle jaws.

■ LAMINATE CURVES

Shopmade jigs allow you to glue curves.

When buying clamps, look closely at the grip and the threads. A small grip is difficult to turn under high clamping loads and may not achieve adequate pressure. Look for a large, comfortable handle and wide, flat-looking Acme threads, which are easier to wind and unwind. No matter how many clamps you buy, you'll still have jobs that use every one you own.

Sanders and Shapers

When joints don't line up the way they should, new woodworkers often grab a sander and lean into the high spot. Peering through the resulting dust cloud, it may seem that things are getting better, but they probably aren't. Sure, the transition at the joint is smooth, but it's not flat—it sits in a hollow dug by the sander. A shiny finish will magnify the flaw.

It's a classic problem that occurs when sanding by hand or machine. You can't get a flat surface when sanding. Sanding is for smoothing surfaces that have already been planed flat and for preserving that flatness by using diligence and the proper technique.

The first tenet of careful sanding tells us to keep flatness in mind. If you're working on a flat surface, keep it that way. Start by applying only light pressure using a coarse abrasive (80 grit or 100 grit for hardwoods, 120 grit for softwoods) on a stiff pad or block—cork or felt blocks are the traditional

WATCH OUT

- Don't push down on a random-orbit sander. Its own weight is enough.
- Keep the sander flat on the surface. Tilting on the sander destroys flatness.
- Step through sandpaper grits from coarsest to finest (to 220# for finishing).
- After machine sanding with your finest grit, touch up with hand sanding.

favorites, but rubber works well, too. Don't linger in any one spot and work the entire surface until it's uniformly scratched. If the surface isn't flat and you have to remove high spots, keep a random motion until the high spot is level to the rest of the surface.

The second tenet of careful sanding says that smoothness is best accomplished by progressing from coarse to fine grits, carefully cleaning the surface to remove the previous

WHAT THESE TOOLS CAN DO

■ SAND FOR FINISHING

Apply light pressure and cover the surface uniformly.

■ POLISH FINISHES, BOATS, AND CARS

6" sanders handle wool bonnets for polishing.

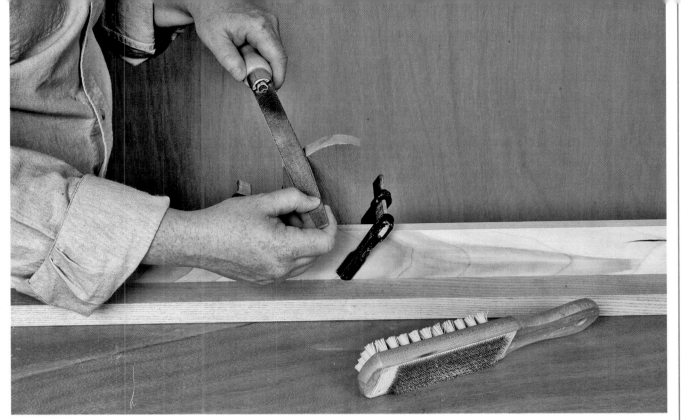

Similar to a file, **but with rows of tiny teeth, a patternmaker's rasp cuts quickly and smoothly. Hold it in both hands as shown and either push or pull it, but don't use it to cut on both strokes.**

grit before progressing. If you're using a machine, cover the entire surface equally and finish off with a few minutes of hand sanding in the grain direction using your finest grit and a block. Sand edges by hand only—sanding an edge with a machine is guaranteed to round it over randomly, marring the crisp line for good.

What to buy

For machine sanding, get a random-orbit sander. The Spirograph® swirl of the sanding disk results from two separate motions—the pad spins in a circle on an eccentric shaft while an offset weight spins a slow elliptical

■ SMOOTH A FLAT SURFACE

A sanding block helps retain flatness.

■ CONTOUR WITH FORMS

Profiled rubber sanding forms aid sanding in tight spots.

Sanding and Shaping Tools

For efficient machine sanding, use a random-orbit sander. You'll also need hand-sanding blocks and a rasp for fast contouring. Use a file to smooth surfaces after rasping, but before sanding.

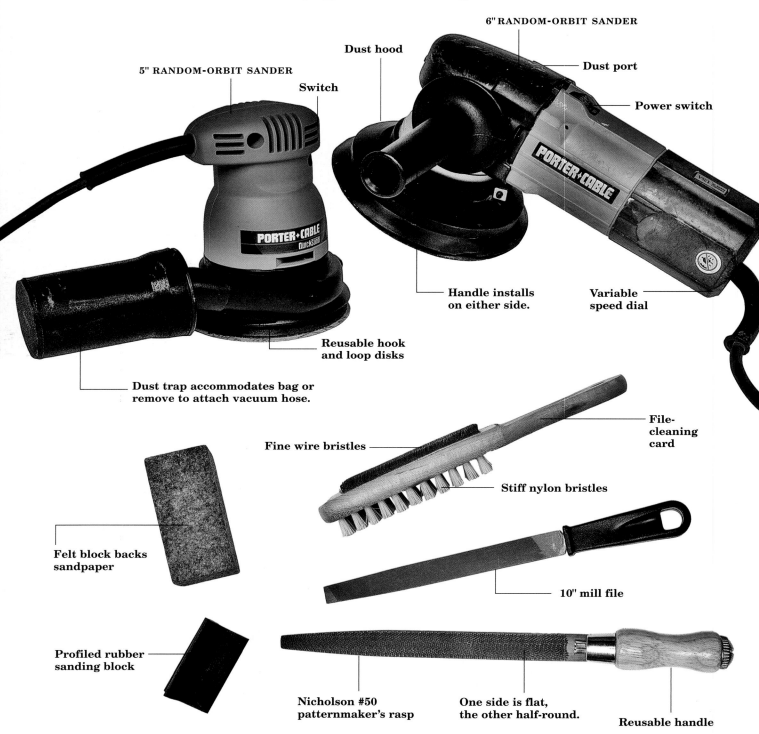

6" RANDOM-ORBIT SANDER

Dust hood

Dust port

5" RANDOM-ORBIT SANDER

Switch

Power switch

Handle installs on either side.

Variable speed dial

Reusable hook and loop disks

Dust trap accommodates bag or remove to attach vacuum hose.

File-cleaning card

Fine wire bristles

Stiff nylon bristles

Felt block backs sandpaper

10" mill file

Profiled rubber sanding block

Nicholson #50 patternmaker's rasp

One side is flat, the other half-round.

Reusable handle

orbit. The action isn't truly random, but if you keep the sander moving, it amounts to the same thing. That means two important advantages over other sanders: It leaves few scratches behind, even when sanding against the grain; and it cuts aggressively without being hard to control.

Sanders with 5"- or 6"-diameter pads sand equally well. The smaller sander is handy inside cabinets and other confined spaces. The bigger sander is a two-handed machine, more comfortable for big jobs. Whichever you choose, look for a variable-speed machine with hook and loop pads, because they can be reused. A dust-collection port is preferable to a dust bag, and a dust hood—a close-fitting, flexible rubber hood between the sander and the surface—helps prevent airborne dust. Also look for a machine that has a pad dampener to slow the motor when it's not actually sanding. With it, you can lift a running sander to check the surface and replace it without gouging the wood.

Choosing and using a rasp

A rasp is an interesting shaping tool that excels at shaping contours and flattening small areas. It's similar to a file, but designed for cutting wood instead of metal. The best rasps have irregular rows of sharp little teeth that cut quickly and smoothly. You'll use a rasp for jobs like rounding corners, shaping end grain, fitting tight joints, and shaping curves.

A trip to a hardware store or home center could turn up five or six different rasps, all of them too coarse for fine woodworking. The only rasp worth buying is a patternmaker's rasp such as the Nicholson® #50. With its close, fine teeth, it's easy to control. Rasps are usually sold without handles; pick up a threaded handle you can reuse.

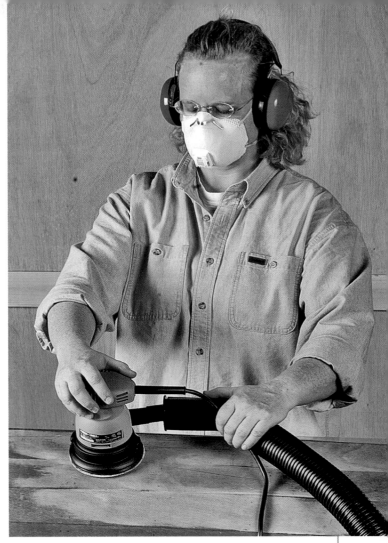

Random-orbit sanding with dust extraction. A random orbit sander is your best bet for an all around sander. It's moderately aggressive but doesn't leave swirl marks. For best results, hook it up to your shop vacuum—it'll run cooler and sandpaper will last longer, not to mention keeping the air clean.

Even a fine-toothed rasp leaves some pretty rough tooth marks. You can smooth them with sandpaper, but you'll get better results if you use a smooth cut mill file first. Though it's really a metalworking tool, it works well on wood.

You'll also want to use the mill file on metal—notably for smoothing the edges of planes and preparing scrapers.

Keep your files and rasps clean with a two-sided brush called a file card. It has aggressive metal bristles on one side, and softer nylon ones on the other.

Shop Vacuum

You need a shop vacuum early on to keep your shop clean and to control dust at its source.

Look for a moderate-size shop vacuum with two-stage filtering, where the large particles drop into a tank while the air exhausts through a separate cartridge filter. Choose a vacuum that lets you add a bag or prefilter between the two. Fitted paper bags neatly dispose waste. A reusable cloth bag or prefilter is less expensive, but emptying the waste puts dust back into the air.

You'll find shop vacuums at two price points—around $100, and $200 and over. The suction power is similar; the difference is in the features. Less costly vacuums often have no way to anchor the 2¼"-diameter hose to the intake port, so you can't lead the vac around by the hose. The motor is also on the top, making the machine both tippy and noisy— around 80db to 90db (wear hearing protection).

For the higher price, you'll get a much quieter machine (57db to 62db). A side-mounted motor provides more stability, and the hose locks into the machine. The pricier machines are designed for both general cleaning and as a dust-extraction system for portable power tools. The hose on these vacuums is often longer and more efficient for dust extraction, but at 1" to 1¼" in diameter, it can be prone to clogging. For better performance when cleaning, get the optional larger diameter hose.

BRUSHES AND BROOMS

A tidy shop is not a moral issue. It simply promotes clear thinking and reduces mistakes and accidents. Keep a soft-bristle bench brush handy to sweep shavings off the benchtop. Keep the floor around your work area uncluttered and sweep up with a soft-bristled upright broom. Remember that sweeping raises dust, so wear a dust mask. A really big dustpan can double as a scoop for picking up shavings and chips, and it stays put so you can sweep directly into it.

Shop Vacuum

Until you're ready for a dust collector, you'll be using your shop vacuum for both cleanup and dust extraction.

Handle for ease of movement

Motor on top

Internal cartridge filter

Optional brush is the most-used tool.

Crevice tool

2 ¼" hose

Floor sweep and extensions

Reducer for using home vac attachments or connecting to hand power tools

Additional length of hose

Hose locks into canister.

Safety Gear

Hazards abound in a woodworking shop, from the obvious to the insidious. To stay safe and healthy, you must pay attention. As a teacher of mine used to say, "Eternal vigilance is the price of safety."

Your overall attitude has a profound effect on safety. Make up your mind to live your shop life at a slower pace and enjoy the process. If you're tired or sick, take a break or do simple maintenance. Take the time to keep a clean shop, and wear safety equipment whenever a hazard is present.

Eyes

With all the dust, chips, and splinters that machine woodworking sends flying, it's not surprising that woodworkers account for thousands of eye injuries each year. Nearly all could have been prevented had the woodworkers worn protective lenses.

Find something that fits right and looks good so there's no excuse not to wear it. Look for polycarbonate lenses that meet the ANSI Z87.140 specification (it's displayed on the packaging). If you wear prescription eyeglasses, their lenses probably aren't designed to withstand much frontal impact, and they don't have side shields. You can clip on side shields or buy goggles to fit over your glasses, but the safest solution is to visit your optician for proper prescription safety glasses.

Ears

Noise damage to your hearing is cumulative and irreversible; woodworking machinery is noisy enough to cause damage. Ergo, you need hearing protection. For intermittent use, earmuffs are your best choice. For long-term use, foam plugs are more comfortable and offer more protection. For the best protection, wear both.

Respiratory system

The dust in shop air is the thing most likely to cause health problems. Woodworkers commonly suffer from asthma, sinusitis, bronchitis, and shortness of breath, and they are more likely to contract a rare form of nasal cancer.

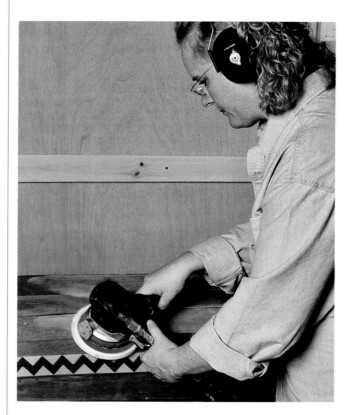

With good dust collection, you can forego a dust mask and special eye protection when using some hand power tools.

Personal Safety Equipment

Working smart and keeping a tidy shop go a long way toward keeping a workplace safe, but some accidents are beyond your control. Guard against surprises by keeping and using all the proper safety gear.

Goggles fit over glasses or can be worn alone.

Seal keeps out dust.

Side shields

Safety glasses

Foam earplugs

Earmuffs

Mask for wood dust

NITRILE
Durable, disposable, nonallergenic, good for most shop solvents

LATEX
Not good for oil-based solvents

COATED COTTON
Nonslip

LEATHER
General purpose, heavy duty

HIGH-TECH CLOTH
Reinforced with KEVLAR® at stress points, general purpose

Wearing it all. Almost all woodworking machines are loud enough to cause hearing damage. They also put out a lot of dust and can project splinters toward your eyes. Be safe: Wear ear protection, dust masks, and eye protection whenever a hazard is present.

It doesn't take much dust to cause problems. The American Conference of Governmental Industrial Hygienists recommends that workers be exposed to no more than a teaspoon of dust in a two-car garage.

The dust you can see in your shop is only part of the problem. The most troublesome dust is the particles that are smaller than 10 microns, and they're invisible to the naked eye. They bypass most of the body's defenses and can enter directly into the sinuses and lungs. They also remain in the air for hours, so be sure to keep your mask on even after the air seems clear.

Disposable masks are fine for dealing with this dust, but be sure to get one labeled for wood dust. You'll get a better fit if it has two bands around the head and a metal strip over the bridge of the nose. Pinch the metal close to your nose to prevent your glasses from fogging.

Hands

Your hands are subject to a variety of dangers when woodworking—sharp tools, splinters, vibration, chemicals, and dirt. Protect them with the right gloves for the job. You'll need a few different kinds:

Leather: These all-purpose gloves protect against splinters and blistering. Cowhide is cheaper, deerskin more supple.

High-tech cloth: With spandex backs and Kevlar® stress points, these lightweight gloves provide comfort and dexterity for just about any job.

Coated cotton: Close fitting and comfortable, these gloves are good for handling sheet goods and surfaced lumber.

Vinyl: Vinyl gloves are readily available and inexpensive, but they tear easily and offer little protection against most solvents.

Latex: These gloves are disposable for working with epoxies, glues, and water-based dye finishes, but they don't stand up well to oil-based solvents.

Nitrile: These provide the best disposable protection from the solvents found in woodworking shops, with the exception of lacquer thinner. They offer good dexterity and more durability, and they are nonallergenic.

Minor cuts, splinters, and bruises are an everyday part of woodworking. Keep a first-aid kit in the shop so you can clean, treat, and cover wounds promptly. The best starting kit is one that meets OSHA standards for the construction industry. They're widely available from pharmacies, industrial suppliers, and medical-supply houses. A two- or three-shelf wall-mounted case is convenient and large enough to house all the necessary items. Few kits will contain all the items below; add the missing ones yourself.

CLEANING AND DISINFECTING

Alcohol cleansing pads
Antiseptic wipes
Providone-iodine solution (Betadine®)
Triple antibiotic ointment
Burn gel or spray
Absorbent gauze compress
Fine tweezers and sterilized
   needles for splinter removal
Eyewash and cup
Saline solution
Cotton swabs

BANDAGES

Assorted adhesive bandages
2" x 4" elbow and knee bandages
Assorted gauze pads
Butterfly closures
Conforming bandage, 2" and 3"
First-aid tape
Triangular bandage/sling
Self-adhesive gauze 1" (for wrapping
   fingers when sanding)
Sterile eye pads

MEDICATIONS

Pain reliever (aspirin, acetaminophen)
Antihistamine tablets (Benadryl®)

OTHER

Instant cold compress
Clean plastic bags
Nonlatex exam gloves
Rehydrating eye drops (GenTeal®)
Small scissors
Krazy Glue® (for small cuts on fingers)

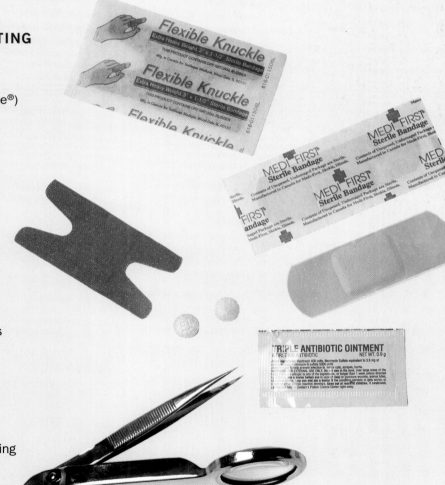

The Essential Shop Space

You'll have to create your first shop wherever you can, most likely in the garage or basement. At this point, you're not ready to devote much space, time, or money to the project, but it's important that you carve out an area dedicated to woodworking (see the floor plan on the facing page). If getting at your bench requires much rearranging or means you have to unfold some clever contraption, you'll soon decide it's not worth the hassle. If you can set aside even a small place that's just for woodworking, you'll be able to go out and work for an hour or two without going through a big production.

Set up your bench and clear some space around it to store your tools, materials, and works in progress. The ideal spot has good access for getting in materials and space for working on big projects—even if you have to shift things around occasionally. You'll also need good ventilation and plenty of light.

Wherever you choose, command that space. Don't make do by merely shoving things aside or working on trash cans or on the floor. Clear out your space and stow things neatly. When you're doing woodworking, stop periodically and put away the tools you're not using at that moment. Take the time to use clamps, sawhorses, and other aides rather than pretending you'll get by without doing it right. Develop these habits from the start, and you'll generate fewer mistakes and have more fun—in both the short and long runs.

GARAGES VS. BASEMENTS

	GARAGES	BASEMENTS
PROS	Good access through overhead doors. Sturdy floors. At-grade access makes it easy to move in machines.	Electrical service usually nearby. Warmth provided by furnace. Sturdy floor. Overhead storage in floor joists.
CONS	May need heat or ventilation added. May need additional power supply.	Potential moisture problems. Low overhead space. Wood dust can infiltrate the house. Below-grade access could make it difficult to move in machinery. Close location to furnace a potential fire hazard for dust and chemicals.

Floor Plan, the Essential Shop

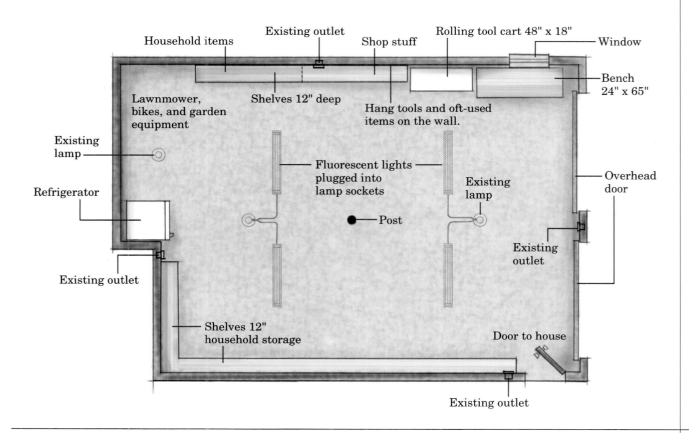

Household items

Existing outlet

Shop stuff

Rolling tool cart 48" x 18"

Window

Lawnmower, bikes, and garden equipment

Shelves 12" deep

Hang tools and oft-used items on the wall.

Bench 24" x 65"

Existing lamp

Fluorescent lights plugged into lamp sockets

Existing lamp

Overhead door

Refrigerator

Post

Existing outlet

Existing outlet

Shelves 12" household storage

Door to house

Existing outlet

If you plan on cutting wood, you need pair of good sawhorses to support the wood and prevent the saw from binding, and to save your bench. You can also lay down a partial sheet of plywood and use the sawhorses as support for an auxiliary bench for layout work or to assemble your projects at a convenient height. Pull one up to the bench for a comfortable seat, or pile them with lumber.

There's no best height for sawhorses. Around 30" high is common, but you could make them higher or lower to suit a particular job. Wide ones are good for dealing with plywood, but they take up a lot of room in the shop—sometimes narrower ones work better.

Sawhorses are another tool for which you shouldn't try to make do. Use good, sturdy horses and clamp your work when necessary.

You can buy your sawhorses ready made, build them from a kit with 2x4s, or build a set to suit. Remember, you never use just one sawhorse; always build them in matched pairs.

Thousands of sawhorse designs are available; take your pick. Just make sure they're sturdy and that you always use them in matched pairs. Shown here from left to right: a shopmade sawhorse, a commercial folding sawhorse with a wooden top, and a knock-down plywood sawhorse that takes up little space when not in use.

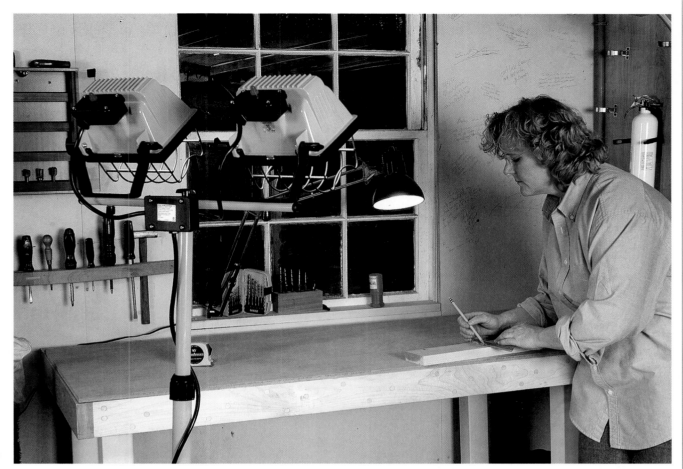

You'll probably have to supplement the lighting in your shop. Use drafting lamps at the bench and add halogen lamps on a stand for bright light where you need it.

Electricity and Lights

If you're lucky, your space will have some three-prong electrical outlets nearby. If not, use extension cords and multiple-outlet surge protectors to get juice where you need it. They don't have to run along the floor—cable-tie them along the wall for a neat semi-permanent solution.

Most basements and garages don't have enough light for woodworking, so add two or three plug-in fluorescent lights to the overhead, as well as a drafting lamp from an office supply store for your bench. Powerful halogen lights on a stand let you direct light where you need it (see the photo above). Later, you'll find they're invaluable for fine finishing work—angle their shine across the surface to highlight irregularities when applying the finish.

Storage

Hang your most-used hand tools on the wall near your bench and keep a section of nearby shelves for boxes, buckets, and tool bags. Most woodworkers try many storage methods and never settle on just one. Store your tools so that they're safe yet handy. The less work you have to do to get a tool (such as unfasten latches or remove trays), the better.

Start your woodworking life with a good mobile cart about 3' tall, 4' long, and 18" deep. Buy it or build it, but have all four casters swivel and at least two lock so it's easy to position the cart where you want it and keep it there.

The Basic Shop

What to Consider

As you gain experience with the tools in your Essential Shop, you'll want to take on more complex projects. Maybe you want to build a bookcase with graceful arches or a sturdy end table with mortise-and-tenon joinery. Perhaps you envision a jewel-like cabinet with dovetailed drawers and hand-carved pulls. The tools in this section make such things possible.

To the Essential Shop's breakdown tool—the circular saw—we've added the miter saw for fast and accurate crosscuts and the jigsaw for cutting curves. With a miter saw, you can take surfaced lumber and break it down into workpieces of nearly any size or shape.

Adding a router to your shop gives you the ability to do all kinds of interesting things to those workpieces. You can cut rabbets, dadoes, and grooves. You can contour edges, mold interesting profiles, and create inlays and other decorative effects. With the right jigs, you can use your router to cut an array of joints, from simple shoulder joints to dovetails and mortise and tenons.

The hand tools in this section allow you to refine the machine work so joints fit better and the finish is finer. And if you're inclined to put in the practice time, you can fashion the same joints with hand tools. The tools in this section are all you need for even the most complex joints.

Hand-tool joinery requires a proper woodworker's bench with a vise or two and a row of dog holes. A bench like this makes working on complex pieces a pleasure, especially if you need to clamp and unclamp frequently to work on a subassembly or to check a joint's fit.

More tools need more space

Your shop will grow to accommodate your higher level of woodworking skill. You'll need more storage space for lumber and tools and more horizontal space to do your work. It

Keep it sharp. Complex projects require sharp tools. To do your best work you'll need to upgrade your sharpening equipment and use it often.

▲ **Keep it organized. You'll need more storage and bench space in your shop to accommodate your new level of skill.**

▶ **Keep it basic. There may be faster ways to do things than by using the tools in this section, but the result won't necessarily be better. Basic tools well handled can produce some very fine work.**

takes some thought to organize the growing complexity, but no great financial investment to have a well-run shop.

You'll be surprised at all you can accomplish with these simple tools. Consider the stunningly complex and perfectly finished furniture of the 17th and 18th centuries (think of Louis the XIV's palace at Versailles and Chippendale). The lion's share of a master cabinetmaker's tools from that era were nothing more than customized measuring and marking tools, planes, chisels, and handsaws.

Basic tools, yes. The results can be anything but.

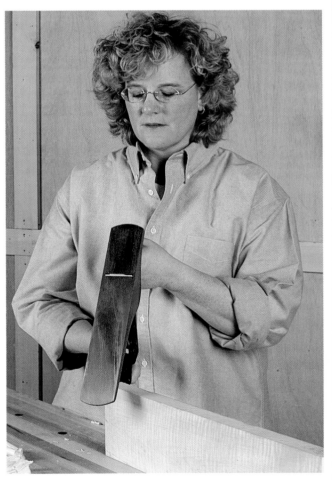

Router

Given a little ingenuity and a good jig, you'll find there's not much a router can't do around the shop. I've seen them used for everything from making dowels and roughing out backgrounds of carvings to profiling elegant edges and cutting intricate joints.

The router is a very simple tool, and therein lies its versatility. It's a motor turning at some 27,000 rpm connected to a bit with a tapered collet that squeezes the bit's shank when a special nut is threaded tight. This unit locks into a handy base that keeps it upright and allows for adjusting the depth of cut. This simple arrangement accepts a multitude of router bits in all shapes and sizes—from 3/16" straight bits to complex bits with diameters of 1½" or more.

Whole books have been written on the subject of routers, and you should study them. Before long, you'll be creating router jigs to solve your own woodworking problems.

What to buy

Start out with a handy midsize router (1½ hp to about 2 hp) with collets to fit bits with either ½" or ¼" shanks (see "What Size Shank?" on p. 55). For maximum versatility,

WHAT A ROUTER CAN DO

■ JOINT

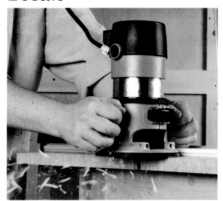

Straighten a wavy edge.

■ PLUNGE

Lower a spinning cutter into a workpiece.

■ PROFILE

Cut a decorative pattern rabbet in an edge.

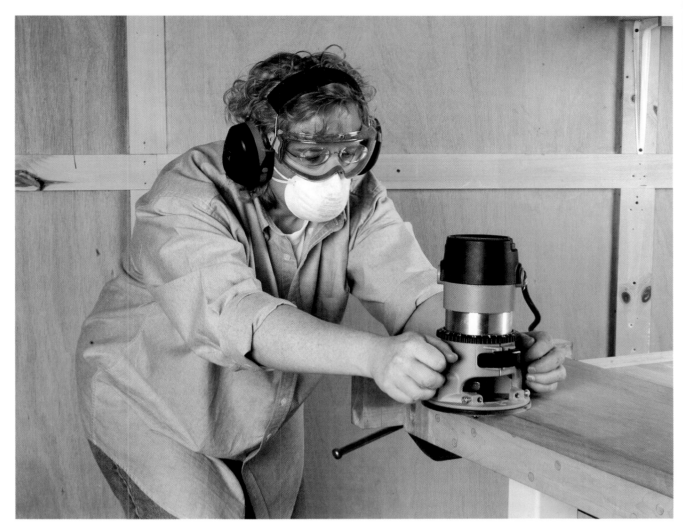

Approach is key. Control the router at all times. Develop a strong stance and get down low enough so you can really see what's going on.

■ **DADO**

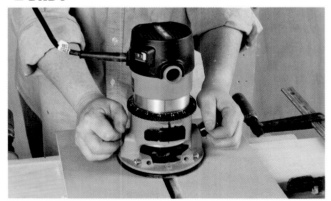

Cut a groove across the width.

■ **CUT TEMPLATES**

Cut an exact copy of a template.

Router with Fixed and Plunge Bases

A router in the 1½-hp to 2-hp range is big enough to do most jobs and not too unwieldy. Be sure to get a router with interchangeable fixed and plunge bases. An optional D-handle matched with an offset base gives a wide footprint ideal for routing edges.

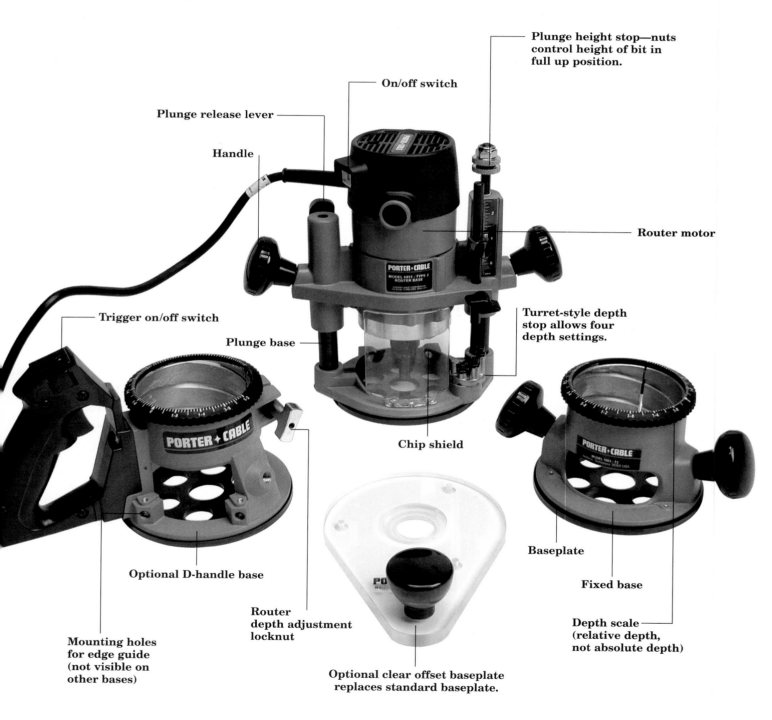

Plunge height stop—nuts control height of bit in full up position.

On/off switch

Plunge release lever

Handle

Router motor

Trigger on/off switch

Turret-style depth stop allows four depth settings.

Plunge base

Chip shield

Optional D-handle base

Baseplate

Router depth adjustment locknut

Fixed base

Mounting holes for edge guide (not visible on other bases)

Optional clear offset baseplate replaces standard baseplate.

Depth scale (relative depth, not absolute depth)

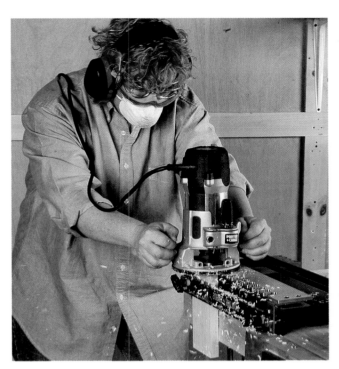

Versatile jigs. You can do almost anything with a router and the right jig. This one routs both parts of a dovetail joint at once. Once it's set up correctly, you can easily rout a kitchen full of identical dovetailed drawers.

Inlaid detail. You can create perfectly fitting inlays of any shape with a router, a template, and the right collars. The inlay will fit perfectly, with just enough space for the glue.

get a kit that includes interchangeable fixed and plunge bases (see "Router with Fixed and Plunge Bases" on the facing page). You'll use the fixed base most of the time because it's lighter and better balanced, but the plunge base lets you safely lower the router bit into the work for routing in the middle of a workpiece. Look for a special sale on a kit that includes both bases—buying bases separately is considerably more expensive.

Your next criteria should be comfort. Find a router that feels right in your hands, one on which the knobs are the right size for you, the on/off switch is conveniently located, and the depth adjustment is easy for you to operate. Check out the plunge base as well and make sure you can operate the plunge release lever easily. And take a look at the depth stop—you'll want one that adjusts to cut at accurate depths, without needless complexity.

Then you can consider the niceties. A spindle lock button holds the arbor from turning so you'll need only one wrench to change bits. A soft-start motor gradually ramps up to speed, eliminating the high-torque twitch of a router going from 0 rpm to 27,000 rpm in an instant. Variable speed lets you reduce the rpm when turning large-diameter bits, a feature you'll find handy from time to time.

Beyond edges

When you first start using a router, you'll mostly use it for dealing with edges—straightening, profiling, or making them more interesting (see "Five Ways to Guide a Router" on p. 54). These everyday tasks are a great way to get familiar with the router, but you'll soon find other uses. One of my favorite router applications is making identical parts from a template. Once you have a good

SHANK-MOUNTED BEARING

A template secured atop the workpiece guides a bearing mounted on the bit's shaft. If the bearing's diameter is equal to the bit diameter, it produces an exact copy of the template.

TIP-MOUNTED BEARING

Tip-mounted bearings usually run against the workpiece's edge or against a template. They're often used for molding decorative profiles, for cutting rabbets, or for use with templates mounted below the workpiece.

COLLAR

A collar screwed in the base plate runs against a template or guide, offsetting the cut from the edge. There's some math involved to position the cut correctly, but collars offer a wider depth adjustment range than bits with bearings.

FENCE

Some router baseplates feature a straight side to run along the fence. If yours doesn't, put a mark on your base and keep that point against the fence.

EDGE GUIDE

A fence attached to the baseplate makes simple work of routing lines parallel to edges. Modify the guide for less contact surface, and it will follow a curved edge.

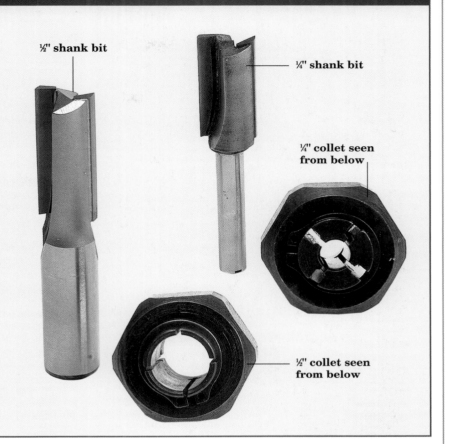

When you set out to buy router bits, you'll find they come with either a 1/4" or 1/2" shank diameter. The 1/4" size is fine for small bits, but use the larger shank diameter whenever you have a choice. It'll give a cleaner cut because it runs with less vibration, and it's less likely to bend or break under load. Although they're offered for sale in some places, avoid 1/4" shank bits with outside diameters of more than 11/2". The shank is too thin for best performance.

A choice of collets. Most routers have interchangeable collets to run bits with either 1/2"- or 1/4"-diameter shanks.

½" shank bit

¼" shank bit

¼" collet seen from below

½" collet seen from below

template, it's easy to run a straight bit with a bearing around the edge for perfect copies.

With the right jigs and bits, routers can solve a multitude of problems. They can cut perfectly fitting inlays—either thin ones for decoration or thicker ones to fill a knot hole or make a structural repair. Mounted on an arm pivoting around the center point, the router is an accurate and secure way to cut a perfect circle. A router can cut the classic joints of hand woodworking—the mortise and tenon and the dovetail (see the top left photo on p. 53)— and it can straighten and smooth as well.

Perfect circles every time. The router is a great way to cut circles. For perfect arcs, mount it on a long arm and pivot it around a nail or screw.

Compound Miter Saw

A miter saw is a specialized tool for dealing with crosscuts, which is no small matter in woodworking. Fine work and well-fitting joints are impossible without smooth ends cut at the correct angle. A miter saw does this quickly and with relentless accuracy.

The blade on a simple miter saw pivots up and down for straight cuts and to the right and left for mitering. Compound miter saws also allow the saw to pivot from side to side for beveling. The maximum cutting width depends on the blade diameter. A 10" blade can crosscut a board 5½" wide, while a 12" blade cuts 8". To cut wider boards, you'll need a sliding miter saw, where the blade arm is mounted on rails, allowing it to slide back and forth. This increases the maximum width to 11½". Some saws offer all these features—they're called sliding compound miter saws.

What to buy

You'll want a lot of versatility from your miter saw, so look for a 10" sliding compound miter saw. It costs twice as much as a non-slider, but you won't regret paying for the additional capacity. You might apply the same logic and step up to a 12" saw, but for most people and most jobs, it's more saw than you

WHAT A MITER SAW CAN DO

■ CROSSCUTS

Cut straight down across the grain.

■ SLIDING CUTS

Crosscut wider boards by sliding the blade out, then down, then back.

need. It's also bigger, heavier, and considerably more expensive.

For maximum versatility, get a saw that can miter more than 45° on one side. Similarly, a dual-bevel saw tilts right and left, a handy feature you'll appreciate from time to time. The table should unlock and pivot easily, and the handle and trigger should fit your hand comfortably. A soft start motor that quickly ramps up to speed reduces the saw's kick when switched on. To keep your hands safe, make it a priority to get a saw with a good clamp for holding down the workpiece at the left side of the blade.

A miter saw is not truly functional out of the box. Before you can use the saw with any degree of safety and accuracy, it needs a proper fence and a system for supporting the overhanging boards.

Your improved fence should be higher and made of a continuous length of plywood. The first cut will make a slot, or kerf in the fence. It's an ideal reference point—line up your mark with the kerf, and you know the cut will be perfect. This new fence also reduces tear out. Anytime you change the cutting angle or the tilt of the saw, you'll make a

Use correct form. When using the miter saw, hold the workpiece on the right side of the blade. With your hand on the left side, a kickback could pull your hand into the blade.

■ **ANGLE CUTS**

Adjust the blade to crosscut at an angle.

■ **COMPOUND ANGLES**

Cut angles with a sliding compound miter saw.

10" Sliding Compound Miter Saw

Miter saws excel at accurate crosscuts, whether they're square, angled, or compound cuts. The head runs back and forth on rails to cut boards up to 11½" wide.

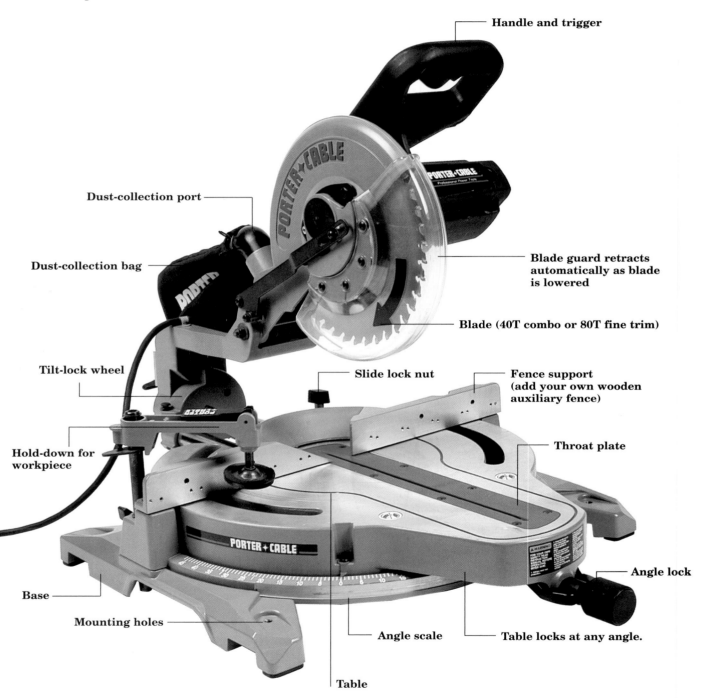

Handle and trigger

Dust-collection port

Dust-collection bag

Blade guard retracts automatically as blade is lowered

Blade (40T combo or 80T fine trim)

Tilt-lock wheel

Slide lock nut

Fence support (add your own wooden auxiliary fence)

Throat plate

Hold-down for workpiece

Angle lock

Base

Mounting holes

Angle scale

Table locks at any angle.

Table

Mobile stand for easy transport. A mobile miter-saw stand makes sense in a small shop. This one has telescoping supports on both sides, and it easily folds and wheels away.

new kerf. Build the improved fence knowing you'll replace it regularly.

Before you can use your miter saw with any degree of safety and accuracy, you'll need a way to support the boards so they won't fall or bind at the end of the cut. If they shift or tilt while the blade is turning, the ends are no longer accurate, not to mention the damage that could result from crashing to the ground. Rig up a way to hold the boards that's easily adjustable for various lengths.

If your saw simply sits on a bench, you can make heavy L-shaped brackets that sit on the benchtop, support the wood and can be moved around as needed. Or put the miter saw in a well so its table is level with the benchtop. Roller stands also work well as supports, or you can buy one of the lightweight adjustable stands favored by trim carpenters (see the photo above).

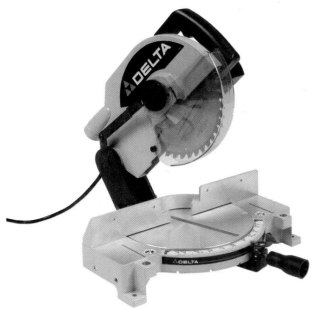

A miter saw that doesn't slide is smaller, lighter, and less expensive than a slider. It also has less cutting capacity. This compact saw is limited to cutting boards less than 5½" wide.

Jigsaw

Using a jigsaw adds a new dimension to your woodworking—curves. From the long arches of an Arts and Crafts–style bookcase to the tight scrolls of a decorative shelf bracket, this saw can handle every type of curve. It also handles straight cuts, and it is one of the safest tools to grab for making quick cuts by eye. Because the blade on a jigsaw moves up and down, as long as you keep the sole on the workpiece and don't twist the blade too much, it won't kick back or otherwise misbehave.

Making a cut with a jigsaw is easy. If the saw is hard to push or steer, something's wrong. The blade may be the wrong type or dull, the cutting speed may be wrong, or the orbital adjustment may be set incorrectly for the type of cut. The orbital adjustment adds a back-and-forth component to the saw's usual up-and-down motion.

The highest orbital setting cuts aggressively, powering through the wood and leaving a ragged edge. Use this setting for rough cuts in solid wood. For smooth cuts, zero out the orbital action. The saw cuts more slowly, but leaves a smoother edge.

Most jigsaws have variable speed. Some build it into a sensitive trigger switch; others use a dial for speed control and a simple on/off switch. In most cases, you won't need to vary the speed as you cut, but you might need to vary the speed to suit the material you're cutting. Use lower speeds for metals, higher speeds for wood.

What to buy

Don't bother buying a jigsaw without orbital action; it's not up to serious work. Make sure you choose one with electronic speed control, and a base tilts in both directions. Have a good look at the blade lock and choose a method that's simple and secure. Some of the "tool-less" methods are so fussy it's easier to use a tool.

WHAT A JIGSAW CAN DO

■ CUT CURVES

Cut freehand curves from flowing to tight.

■ CUT ANGLES

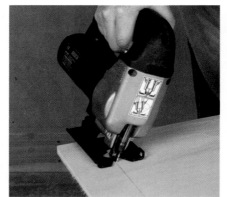

Tilt the blade to the right or left.

■ CUT SCROLLS

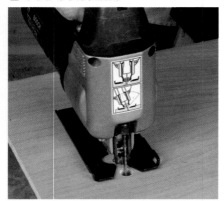

Start your cuts in a drilled hole.

Orbital Jigsaw and Jigsaw Blades for Woodworking

Look for a machine that has adjustable orbital action and variable speed. Triggers that lock easily into the "on" position are useful, and a tilting base allows you to make cuts at angles other than 90°.

The right blade is crucial for best performance. Change blades often—that's why they come in packs of five.

Blade lock (this saw uses a screwdriver to turn a screw)

Variable speed trigger

Button locks trigger "on"

Allen wrench for tilting base

Electronic speed control

Slots for fitting edge guide

Air blower adjustment for clearing sawdust

Blade guide

Orbital adjustment

Not shown—plastic shoe for sawing materials that scratch easily

(6 TPI) Use for rough cuts

■ CUT STRAIGHT

Use a straightedge or guide.

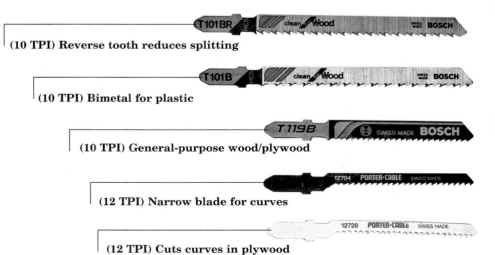

(10 TPI) Reverse tooth reduces splitting

(10 TPI) Bimetal for plastic

(10 TPI) General-purpose wood/plywood

(12 TPI) Narrow blade for curves

(12 TPI) Cuts curves in plywood

Bench Planes

Many modern woodworkers ignore the bench plane under the illusion that power-driven tools are somehow better. Compared to a plane, these crude tools don't do most jobs as well or as easily. Once you know your planes, it's a snap to set one to remove superfine shavings—as thin as four ten-thousandths of an inch thick—to sneak up on the fit of a joint or to smooth rough swirls of grain.

What to buy

First get a 9" #4 or #4½ smooth plane. I like the wider #4½ because it uses the same irons as the #7. This smaller plane is best for smoothing surfaces after they're flat and for flattening smaller workpieces.

Then get a #7 jointer plane, about 22" long. For flattening, it's the best tool in your shop. You should use it whenever the workpiece is large enough to support it.

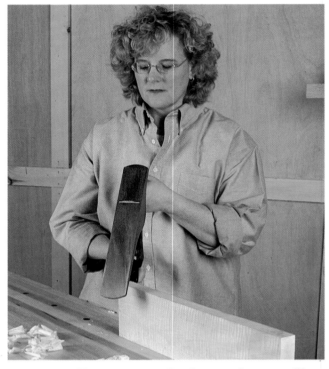

Tuning up. To get top results from a plane, you'll have to spend some time practicing with it, getting to know its ways, and making sure it's sharp and properly adjusted.

WHAT BENCH PLANES CAN DO

■ **JOINT**

Flatten and square an edge.

■ **SMOOTH**

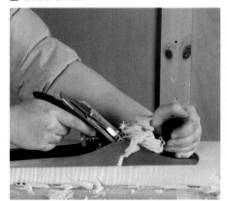

Make a surface smooth while maintaining flatness.

■ **FLATTEN**

Remove bumps and hollows to make a surface level.

Bench Planes

These three planes will handle most of your planing jobs, from rough to supersmooth finishing.

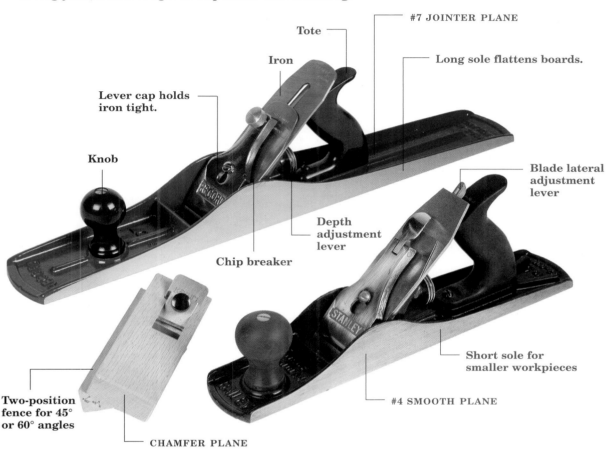

Tote

Iron

#7 JOINTER PLANE

Long sole flattens boards.

Lever cap holds iron tight.

Knob

Blade lateral adjustment lever

Depth adjustment lever

Chip breaker

Short sole for smaller workpieces

#4 SMOOTH PLANE

Two-position fence for 45° or 60° angles

CHAMFER PLANE

■ CHAMFER

Cut a bevel along an edge.

You'll have to look a little to find these planes—most places sell only the #5 jack plane, supposedly the "jack of all trades." It's really too short for jointing and too long for anything else. It's better to start with two planes and let each do the job it does best.

You can do all the chamfering you need with these two planes and your block plane, but a dedicated chamfer plane ensures that your chamfers are all at the same angle and depth, something you'll notice on long edges.

More Chisels

At some point in your growth as a woodworker, you'll find that carpenter's butt chisels are too short to work inside cabinets or drawers, and their blades are too thick to fit in tight joints. You need a set of cabinetmaker's bevel-edge bench chisels. With their longer blades and handles, they allow more leverage. The deeply beveled edges let them slip into close corners, and their fine balance feels good in your hands.

Your basic choice is between Japanese or European styles. They differ on two levels—shape and steel. Japanese-style chisels are a bit shorter with a slight curve in the handle, and the backs have a hollow in the center. A hard steel edge forged to a softer iron body acts as a shock absorber to preserve the edge. Most European-style chisel blades are ground out of a homogeneous blank of machine-forged steel. A slight majority of expert woodworkers favor the Japanese style, but in the everyday world, both do the job.

THE EUROPEAN JOINER'S MALLET

The wooden joiner's mallet is an ancient tool that developed over centuries into the best tool for striking chisels. It's not as hard on wooden-handled tools as a hammer, and its natural resilience is easier on your wrists and elbows. The distinctive keystone-shaped head presents maximum surface area for striking tools. The flat face also concentrates the blows in one place so each blow is as efficient as it can be.

WHAT CHISELS CAN DO

■ FLATTEN

Flatten a small area with the back down.

■ CLEAN A GROOVE

Remove tool marks and irregularities.

■ FIT JOINTS

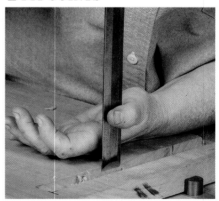

Slice away irregularities and high spots.

More Chisels

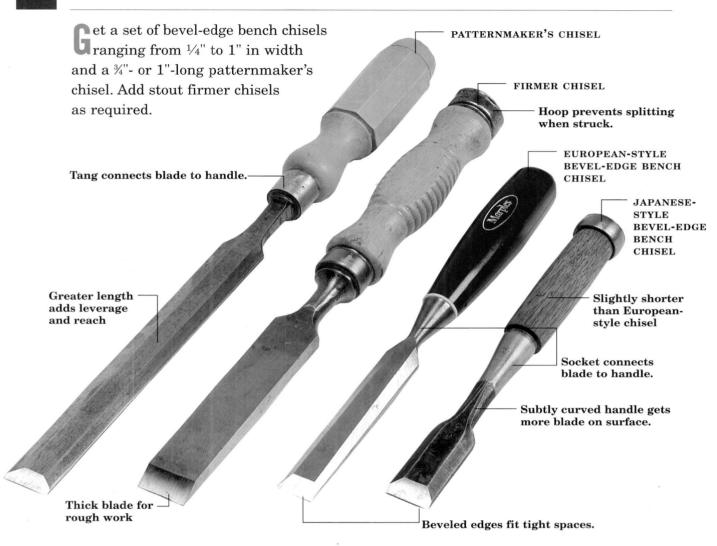

Get a set of bevel-edge bench chisels ranging from ¼" to 1" in width and a ¾"- or 1"-long patternmaker's chisel. Add stout firmer chisels as required.

PATTERNMAKER'S CHISEL

FIRMER CHISEL

Hoop prevents splitting when struck.

EUROPEAN-STYLE BEVEL-EDGE BENCH CHISEL

JAPANESE-STYLE BEVEL-EDGE BENCH CHISEL

Tang connects blade to handle.

Slightly shorter than European-style chisel

Greater length adds leverage and reach

Socket connects blade to handle.

Subtly curved handle gets more blade on surface.

Thick blade for rough work

Beveled edges fit tight spaces.

Round or chamfer corners and edges.

Besides the bevel-edge chisels, you'll need a few stout firmer chisels for heavy-duty jobs. Buy a set if you find a deal; otherwise buy them as needed.

You'll also want a ¾"- or 1"-long thin patternmaker's chisel. The extra length adds leverage and lets you use the chisel bevel-up farther in from the edge of a board. A patternmaker's chisel is for paring and is never struck.

Sharpening Tools

It's been said that civilization began with learning to sharpen; improving the odds of the hunt led to good health and allowed enough free time to develop art. You can expect similar results in your woodworking. Until you get sharpening down cold, you'll be struggling too much to have fun or get beautiful results.

The sandpaper sharpening method described on p. 21 is a great way to get started sharpening, but when you begin building more complex pieces, you'll need the best edges you can get (see "Five Steps to a Perfect Edge" on p. 69). Sandpaper is too coarse to produce the finest edge.

What to buy

The best way to flatten the backs of edge tools is with a coarse (about 220 grit) stone. I prefer diamond stones because they're so hard they cut the fastest. Plus, diamond stones don't dish with use and stay so flat you can use them for flattening other types of

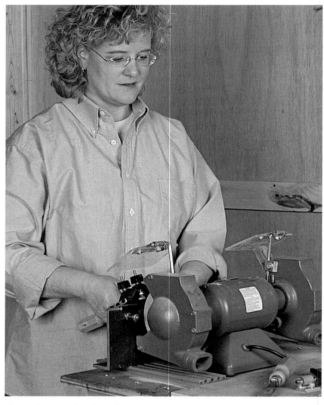

A task worth the time. Truly sharp tools require investments of your time and money. The payoff is high—better results, less effort, and a lot more fun.

WHAT THESE TOOLS CAN DO

■ FLATTEN BACKS

Flatten the backs of planes, chisels, and other cutting tools.

■ GRIND

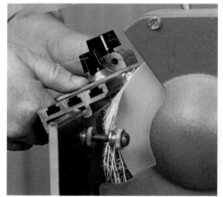

Rough out the bevel at 25° for faster honing.

■ HONE

Refine the edge at 30° to sharpness.

Tools for Flattening and Honing

I t takes a lot of gear to get the sharpest edge, but as always, the right tools make all the difference.

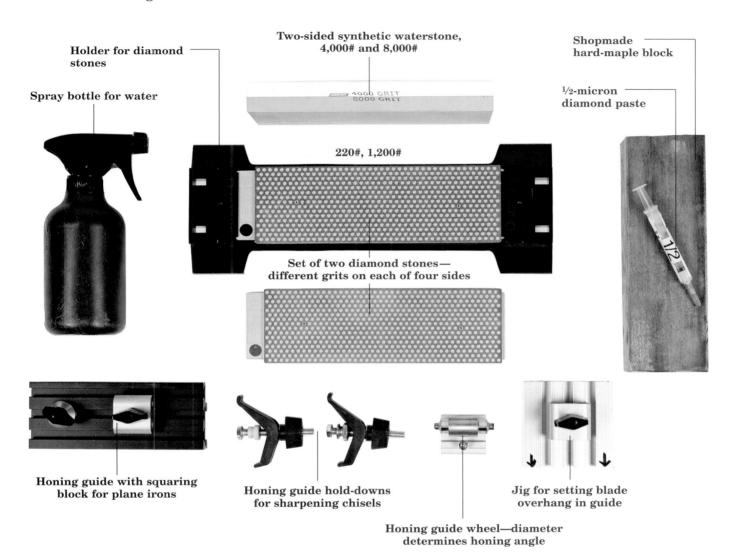

Holder for diamond stones

Spray bottle for water

Two-sided synthetic waterstone, 4,000# and 8,000#

Shopmade hard-maple block

½-micron diamond paste

220#, 1,200#

Set of two diamond stones— different grits on each of four sides

Honing guide with squaring block for plane irons

Honing guide hold-downs for sharpening chisels

Jig for setting blade overhang in guide

Honing guide wheel—diameter determines honing angle

stones, plane soles, and more. Whatever type of stone you get, you'll need three or four stones ranging from 220 to 1,200 grit.

Your first honing guide might not be wide enough to accommodate a #7 plane iron— now's the time to upgrade. You'll also need a guide for your grinder. You can buy the two items individually or get a modular system that uses the same parts for both.

Get a slow-speed grinder (1,800 rpm or so) and a soft wheel for sharpening. This combination runs cooler and reduces the chances of ruining the blade's temper from overheating.

Synthetic waterstones give the best edge, and you'll need at least two grits. Depending on the brand, the coarser grit should be 3,000 to 4,000# and the finer 6,000 to 8,000#. Get a single two-sided stone, or two separate stones.

The Slow-Speed Grinder

Aslow-turning grinder with a soft wheel runs cooler and keeps your blades from overheating and losing temper. Either a 6" or 8" wheel is fine, though the 6" wheel has a lower velocity at the rim.

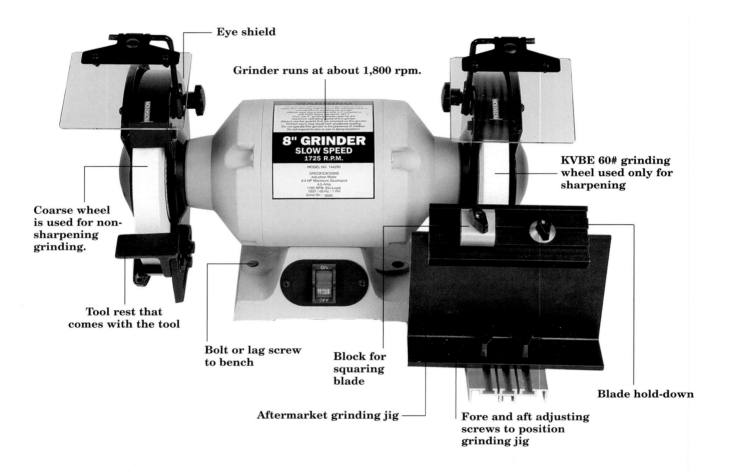

Eye shield

Grinder runs at about 1,800 rpm.

8" GRINDER
SLOW SPEED
1725 R.P.M.

KVBE 60# grinding wheel used only for sharpening

Coarse wheel is used for non-sharpening grinding.

Tool rest that comes with the tool

Bolt or lag screw to bench

Block for squaring blade

Blade hold-down

Aftermarket grinding jig

Fore and aft adjusting screws to position grinding jig

Finally, for the ultimate edge, you'll need some diamond paste and a little mineral oil to cut it. Since diamond paste is usually sold in kits for gem polishing and electronics, you'll probably end up with four or five grits from 9 microns to ½ micron. You only need the finest for honing, but you can use the others to polish the backs of your blades. Use the diamond paste on thick maple blocks cut to the same size as your sharpening stones.

Flat stones are crucial to a good edge. Flatten them often on your diamond stones.

FLATTEN AND POLISH THE BACK

A perfect edge starts with flattening the back on a fast-cutting coarse diamond stone and progressing to your finest grit stones. For the ultimate mirrorlike polish, finish with ½-micron diamond paste on a flat, smooth block of hard maple.

GRIND AT 25° FOR FASTER HONING

Grind the bevel to 25°. Later, when you hone at 30º, you'll remove material from only the tip of the tool, which takes just minutes. Grind again after four or five honing sessions.

HONE THE BEVEL AT 30°

Start honing with your second-finest stone for about a minute. Then create a slight crown in the edge to prevent the corners from digging. Press a little harder on each corner for several strokes.

POLISH THE BEVEL

Using your finest stone, polish for about a minute, pressing on alternate edges to maintain the crown. Then work the back for about 15 seconds. For ultimate sharpness, polish both sides with ½-micron diamond paste.

TEST THE EDGE

Hold the blade loosely in one hand and gently touch the edge to a thumbnail. A sharp edge bites in, a dull one slips or scrapes. If the iron doesn't pass this test, spend more time working the back on your finest stone.

Measuring and Marking Tools

When you're first learning woodworking, your hand skills (or lack thereof) are your roadblocks to success. Once you understand how to use the tools, a new roadblock arises: your ability to measure and mark correctly. Accurate layouts are mostly a matter of patience, but good tools play an important role.

Moreover, there's a completely new set of skills you need to learn in order to measure properly—things like how to present a square to an edge so it gives a true reading and how to use a knife to mark a line in a way that won't damage your straightedge or the blade. When you reach this level of woodworking, a little bit off is too much.

What to buy

If you're serious about doing good work, keep a couple of sliding squares close by at all times (see "The Indispensable Sliding

Quality tools for quality work. Good measuring tools are essential for fine joinery. Use them to check your handwork often, and correct small issues before they become larger problems.

WHAT THESE TOOLS CAN DO

■ LAY OUT 45° ANGLES

Use the 45° leg of a combination square to mark an angle.

■ MARK CUT LINES

Use a knife to scribe lines for accurate joinery.

■ MARK ANGLES

Use a sliding bevel that adjusts to any angle.

Measuring and Marking Tools

Get good tools and use them wisely to ensure that all your joinery comes out square and accurate.

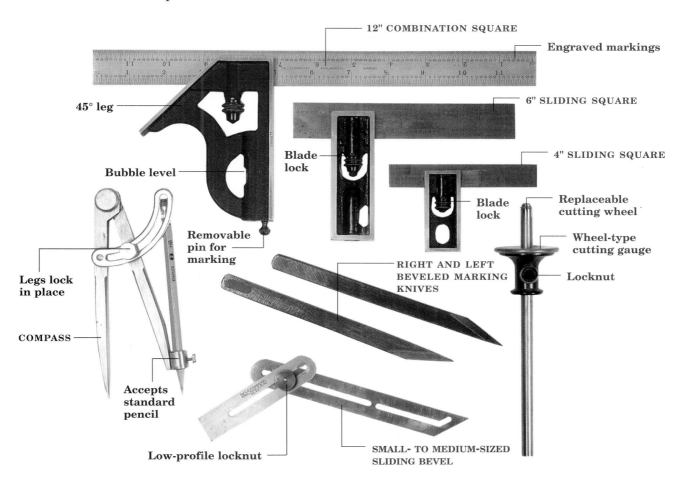

12" COMBINATION SQUARE

Engraved markings

45° leg

6" SLIDING SQUARE

Blade lock

4" SLIDING SQUARE

Bubble level

Blade lock

Replaceable cutting wheel

Removable pin for marking

Wheel-type cutting gauge

Legs lock in place

RIGHT AND LEFT BEVELED MARKING KNIVES

Locknut

COMPASS

Accepts standard pencil

Low-profile locknut

SMALL- TO MEDIUM-SIZED SLIDING BEVEL

■ SCRIBE A LINE PARALLEL TO AN EDGE

Use a cutting gauge with a fixed knife to scribe a line a set distance from an edge.

■ CHECK SQUARENESS

Make sure the edge is parallel to the side.

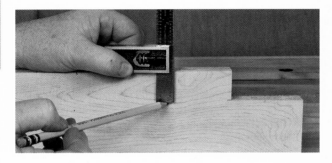

TRANSFER LINES
Rather than use a tape measure or ruler to locate a line, put the square on the edge and slide the blade to the line. Lock the blade in place and transfer the measurement without having to read the scale.

SQUARE AROUND A BOARD
Put the point of your pencil on the mark and slide the square up to it. Hold the square firmly and draw the line. Rotate the workpiece and repeat the process right around the board.

CHECK ANGLES
By changing the blade position to just below flush, you can check inside and outside corners for squareness.

MARK A LINE PARALLEL TO AN EDGE
Using the base as a reference point, slide the scale to the desired measurement. Place a pencil firmly against the end of the blade and push the square along one edge.

MEASURE DEPTH
Slide the blade down into the recess and lock it in place. Either read the scale or transfer the measurement directly.

Though used mainly for laying out and marking joints, your measuring tools have other applications. You can use them to check that your hand power tools are set up correctly—for example, you can check the jigsaw's tilt against the sliding bevel and use a square to true up the position of the grinder jig so it's square to the wheel and the right distance from its face.

You should also use the tools for developing your hand skills. Practice with your bench plane on some fragrant pine, using the square to check for perpendicularity at short intervals. Learn how to hold your plane and distribute your weight until the square shows no changes. Set the sliding bevel to a random angle and plane the angle, then plane back to square.

Square" on the facing page). I'm hard-pressed to say which I use most—a 4" or a 6" square. Get the one that appeals to you most, then ask for the other for your next birthday. You'll find you use a 12" combination square less often, for larger measurements and for laying out 45° angles.

Small- to medium-sized sliding bevels (6" or less) are more comfortable to use than large ones. I like the precision of an all-metal tool, but whatever you buy, remember that large locking nuts get in the way. You'll also need a pair of marking knives, beveled right and left so you always put the flat back against the straightedge. You'll also need a cutting gauge for making knife cuts on face grain and endgrain at a set distance in from the edge of a board. The wheel-type cutting gauge is a real advance from traditional cutting and marking gauges. Finally, seek out a durable compass that accepts a regular pencil and locks the legs in place so they won't shift in use.

A pair of winding sticks that are perfectly straight, square, and true can help you see twist in a board. Sight across the top of the near stick to the top of the far one. If the board is flat, their tops are parallel.

Handsaws

No matter how many power saws you own, there's always a place for hand-saws. With a little practice, they afford a level of control you can't achieve with a power saw. And you'll be surprised how often it's easier to grab a crosscut saw to cut a board to rough length than it is to get out an extension cord and set up a power saw.

What to buy

Get an aggressive saw for rough cutting to length. Any hardware store or home center offers at least a couple of choices, usually a traditional European-style saw, and a hybrid Japanese-style saw (see "Push or Pull?" on the facing page).

Next, get one or more backsaws—fine-toothed saws with reinforced backs for stiffness. These are tools you'll need for hand-cutting joints like dovetails and mortise and tenons. Your first choice should be one 8" to 10" long, either European or Japanese style.

Success with a handsaw is all about technique. Align your body to the task and give yourself plenty of room. With a little practice, you'll find that effortless rhythm that means everything's right.

WHAT HANDSAWS CAN DO

■ ROUGH-CUT LUMBER

A handsaw is faster than a power saw for a cut or two.

■ CUT CURVES

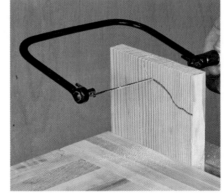

A thin, fine-toothed blade will cut almost any shape.

■ CUT JOINTS

A stiffened blade is best for cutting accurate joints.

Choose Your Style

Start your saw collection with a rough crosscut saw, a backsaw (either European or Japanese style), and a coping saw.

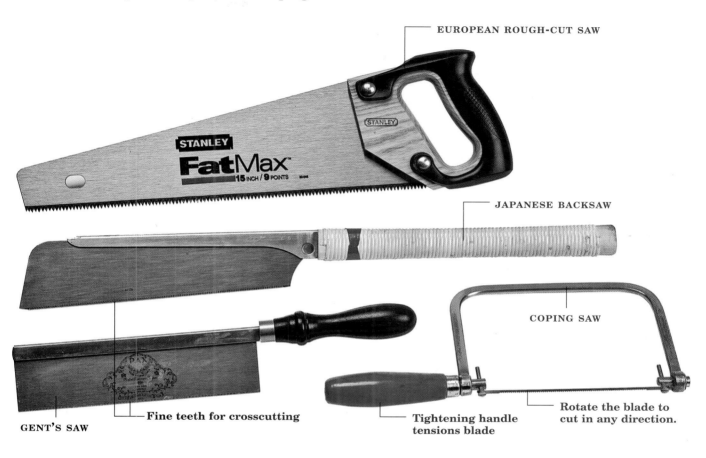

EUROPEAN ROUGH-CUT SAW

STANLEY

FatMax™ 15 INCH / 9 POINTS

JAPANESE BACKSAW

COPING SAW

GENT'S SAW

Fine teeth for crosscutting

Tightening handle tensions blade

Rotate the blade to cut in any direction.

PUSH OR PULL?

The essential difference between European and Japanese saws is in the teeth. A European backsaw cuts on the push, which is handy for very fine work because the sawdust accumulates at the back and doesn't obscure the view. The teeth on a Japanese backsaw are thinner and cut on the pull stroke. Many beginning woodworkers prefer Japanese saws because they cut fast and clean, and they are easy to control.

Within the European style you'll have a choice of the round-handled gent's saw or the more familiar looking dovetail saw. If you make a lot of tenons by hand, you'll also want to get a longer tenoning saw with its unique fine teeth set for ripping.

Then pick up a coping saw. It's the best tool for cutting complex compound curves. One great feature of these saws is that you can rotate the blade to saw at the most advantageous angle. Get a selection of blades—coarse, medium, and fine—to handle whatever job arises.

Scrapers

Scrapers are versatile tools suitable for extremely different scraping tasks. They can cut thin, planelike shavings for flattening and smoothing the wild grain patterns so often torn out by handplanes. They're also the best way to flatten and smooth thin veneers and inlays. At the other extreme, you can put an aggressive edge on a scraper and use it to remove paint or glue that would quickly dull or even chip a plane blade. The secret to the scraper's versatility is its edge—a hook-like burr that slices wood and immediately curls it into a shaving.

What to buy

You'll want to get three types of scrapers to handle the extremes:

Hand scrapers are nothing more than a piece of hard steel with a burnished edge you push or pull over the surface, changing the angle to control the cut. Hand scrapers come

Comfortable handles and a wide sole make the cabinet scraper a fast way to smooth large surfaces that can't be planed. Unlike a sanding machine, it's quiet and dust-free.

WHAT SCRAPERS CAN DO

■ SMOOTH DIFFICULT GRAIN

Hand scrapers smooth veneers and swirling grain without tearout.

■ FINAL FINISHING

A properly scraped finish doesn't need sanding.

■ REMOVE FINISHES

A triangular scraper levels drips or removes a finish entirely.

An Array of Choices

Whether you're scraping dried glue from panels or smoothing surfaces for finishing, there's a scraper that will do the job.

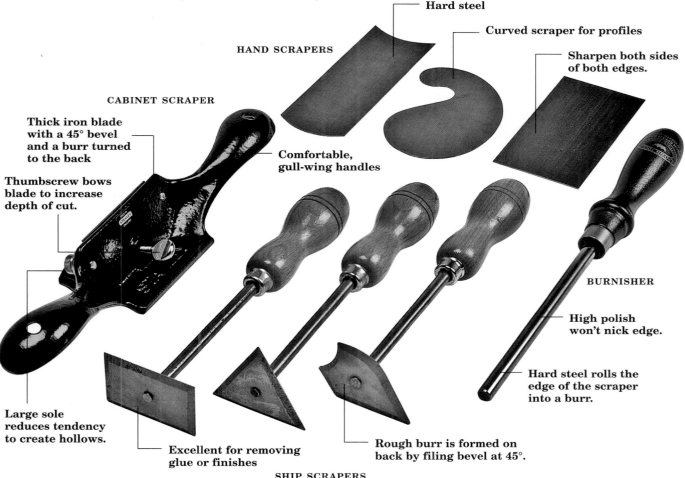

HAND SCRAPERS

Hard steel

Curved scraper for profiles

Sharpen both sides of both edges.

CABINET SCRAPER

Thick iron blade with a 45° bevel and a burr turned to the back

Thumbscrew bows blade to increase depth of cut.

Comfortable, gull-wing handles

BURNISHER

High polish won't nick edge.

Hard steel rolls the edge of the scraper into a burr.

Large sole reduces tendency to create hollows.

Excellent for removing glue or finishes

Rough burr is formed on back by filing bevel at 45°.

SHIP SCRAPERS

in a variety of sizes and shapes, some with curved edges to handle moldings and profiles.

Cabinet scrapers feature a flat sole, comfortable handles, and an iron ground with a 45º bevel that's burnished rather than sharpened on a stone. They're best for scraping large surfaces.

Ship scrapers usually have rectangular or triangular blades and long handles, but also have teardrop-shaped blades for moldings and profiles. They're aggressive tools, best used for removing finishes and glue.

Form a scraper's cutting burr by bending the thin metal edge back into a hook using a hard, smooth burnisher. If a scraper makes dust rather than shavings, it's not sharp.

Woodworker's Bench

Two things distinguish a woodworker's bench: It has at least one vise, and it has at least one row of dog holes (either square or round) along the front edge. The presence of these two items transforms a table into a big, versatile clamp.

Most traditional woodworker's benches have two vises: one set on the front of the bench near the left side (for right-handers) for most simple clamping operations, and a tail vise on the right end of the bench that's used in connection with the dog holes. A wood or metal pin (the dog) set into one of the dog holes on the left end of a workpiece acts as a stop for planing. Put a dog in the hole on the cheek of the tail vise, and you can crank the vise to squeeze the workpiece between the two dogs for a more secure hold.

A woodworker's bench is optimized for using hand tools like planes, chisels, and saws. It offers a variety of ways to hold and support the work at hand.

WHAT A WOODWORKER'S BENCH CAN DO

■ HOLD PIECES IN A VISE

A vise holds the wood securely for a variety of operations.

■ SECURE WORK WITH DOGS

One dog acts as a stop; a second set in the cheek of a vise secures work.

■ HOLD CLAMPS

Secure awkward pieces with a combination of vise and clamps.

A Bench That Holds Your Work

Built on the base of the bench used in the Essential Shop, this bench has all the features you need for advanced woodworking. This uncommon design (suitable for right-handers) uses an iron quick-action vise on the left end that functions as both front vise and tail vise.

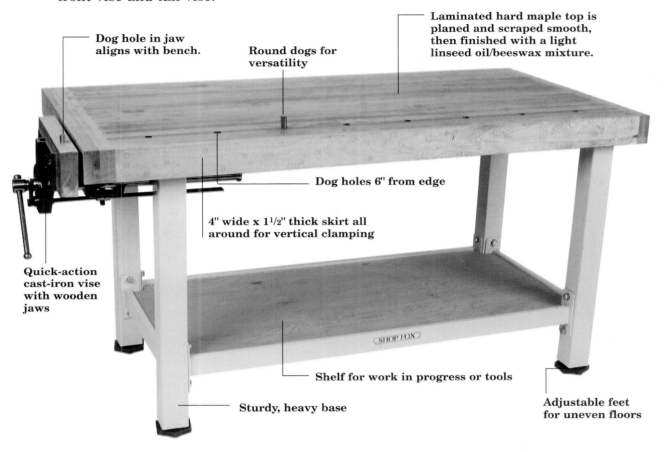

Dog hole in jaw aligns with bench.

Round dogs for versatility

Laminated hard maple top is planed and scraped smooth, then finished with a light linseed oil/beeswax mixture.

Dog holes 6" from edge

4" wide x 1½" thick skirt all around for vertical clamping

Quick-action cast-iron vise with wooden jaws

Shelf for work in progress or tools

Sturdy, heavy base

Adjustable feet for uneven floors

■ SUPPORT BOARDS FOR PLANING

A wooden el clamped in the vise holds the board in place for easy clamping across the bench top.

The bench shown here makes use of a versatile metal vise on the left end of the bench to fulfill the functions of both a front and tail vise (see the sidebar on p. 80). You can stand at the end of the bench when you need the vise to hold small objects, and work along the front of the bench when working with long boards or planing.

What to buy

Whether you build a woodworker's bench or buy one, it should meet the size and height criteria of the Essential Shop workbench (see pp. 8–11). Make sure the rods at the bottom of the jaw are at least 4" below the surface of the bench—if not, you'll find it difficult to clamp wide boards securely. Look for dog holes 4" to 8" in from the edge. Traditional dogs are square, but round ones are more versatile (see "Best Bench Accessories" on the facing page).

Most woodworker's benches have a skirt around the edges; make sure it's at least 1½" thick to provide adequate footing for vertical clamping with big clamps. Look for minimal obstructions beneath the bench so you can clamp across the underside.

Though in-bench storage seems like a good idea, bench drawers are overrated.

They're not really that convenient for holding your tools, and they tend to fill up with dust and shavings. Plus, a few drawers full of tools can add enough weight to a bench to make it difficult for one person to move around the shop. Though you'll store your bench against the wall and often use it in that position, you'll frequently want to pull it out to the center of the shop for access to all four sides. If your floor is uneven, you'll want large, sturdy self-leveling feet.

A light finish seals the benchtop against moisture and makes it easier to clean. An oil-wax or oil-varnish finish works best and is easy to renew.

THE ADVANTAGES OF A QUICK-RELEASE CAST-IRON VISE

Old-fashioned wooden-jaw vises certainly look nice on a bench, but I prefer the utilitarian heft of a big cast-iron vise. First of all, they're easy to install using lag screws or bolts. A few inches of heavy, noncompressible blocking between the vise and the bench get the rods well below the surface of the bench for easy clamping of wide workpieces. The polished metal handle just feels good in the hands, is smaller, and doesn't get in the way as much as the massive wooden ones on an old-fashioned vise.

But the best feature of all is the quick-release lever on the lower right side of the cheek. Put your palm on the center of the T handle, and whether you're right- or left-handed, it's easy to squeeze the lever upward with your finger or thumb as your hand tightens to disengage the threads. Now the vise slides in and out so you can rapidly position your work and tighten the jaws.

CLEVER DOGS

If you go with round dog holes, you have lots of choices when it comes to dogs: short ones, tall ones, plastic ones, and even ones with threaded jaws. Bore another row or two of dog holes in your bench, and you can secure odd-shaped or even curved pieces.

BENCHHOOK

Because of the benchhook's unique Z shape, you can secure a workpiece to it for sawing with one hand. Simply hold the work against the back edge of the hook with your thumb in front and your fingers in back. Stiffen your arm and push. The lip pressing against the front of the bench keeps everything stable so you can saw.

HOLDFASTS

Holdfasts are a great way for supporting a workpiece in the middle of the bench. The old-fashioned one on the left wedges into place when struck. Turn the big brass knob on the holdfast to the right to lock it down.

COVER

Protect your bench with a ¼" Masonite cover. It's much easier to clean than your carefully planed and scraped benchtop, and it's readily replaceable. Put it down whenever you use glue or finishing products and anytime you use your bench for nonwoodworking tasks like fixing the lawnmower.

More Clamps

Woodworking uses an astonishing number of clamps; no matter how many you have, you'll always need more. Keep buying the basics and add specialized clamps as needed.

The new parallel-jaw clamps have a refined head design that ensures the jaws are perpendicular to the bar under load. They are easier to set up, are less likely to twist or mar your work, and clamp more securely than other bar clamps. The only drawback is their high cost, so you may have to build your collection slowly.

Spring clamps are at the other extreme—they're inexpensive and not very powerful. Still, they're great for small workpieces with parallel or nearly parallel sides. A band clamp comes in handy when nothing else will work and can be especially useful when working with curved surfaces.

Corner clamps hold the parts in alignment and free your hands for other things.

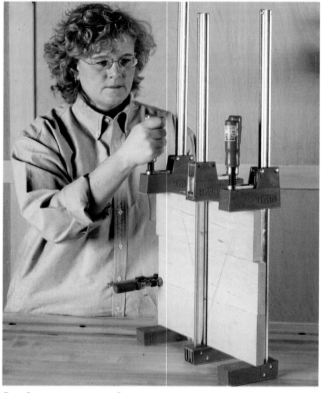

In clamps you need both number and variety. Parallel jaw clamps reduce bouncing while a small clamp on the joint ensures alignment.

WHAT CLAMPS CAN DO

■ SPRING CLAMPS

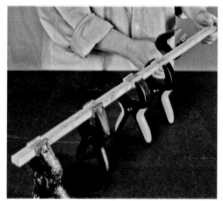

Spring clamps are ideal for small clamping jobs.

■ CORNER CLAMPS

Corner clamps hold parts during fitting and glue-up.

■ WEBBING CLAMPS

Webbing clamps exert even pressure when other clamps can't.

More Clamps

Keep adding to your core collection of clamps and start rounding it out with specialized clamps for jobs large and small.

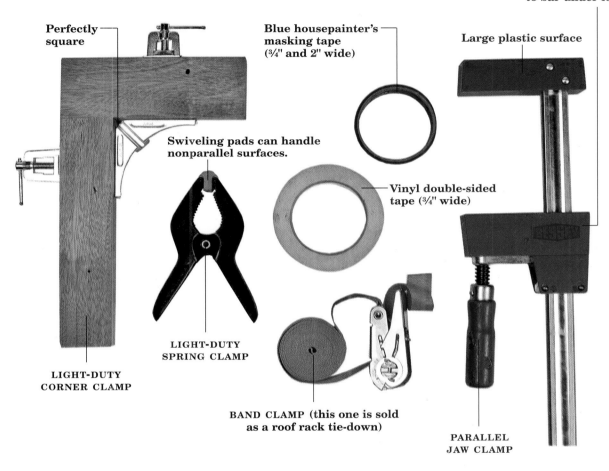

Perfectly square

Swiveling pads can handle nonparallel surfaces.

LIGHT-DUTY CORNER CLAMP

LIGHT-DUTY SPRING CLAMP

Blue housepainter's masking tape (¾" and 2" wide)

Vinyl double-sided tape (¾" wide)

BAND CLAMP (this one is sold as a roof rack tie-down)

Jaws stay perpendicular to bar under load.

Large plastic surface

PARALLEL JAW CLAMP

■ PARALLEL-JAW CLAMPS

These are a handy choice for flat panels and gluing up a carcase.

Start your collection with several small, inexpensive clamps for general duty and plan to buy heavy-duty specialized versions when you need them.

Don't overlook tape. It's great for edge-banding, veneering, and other light-duty work. Double-sided tape is a huge time-saver, useful when holding multiple pieces together and in other cases where clamping would be awkward.

The Basic Shop Space

Once you commit yourself to woodworking, your relationship to your shop changes. You're spending more time there and building more complex projects, so having to move the bikes, reorganize the garden tools, and set up lights before you can work is no longer tolerable. Your growing tool collection is taking up more room, and there's lumber to store, too (see the floor plan on the facing page).

You'd like more room, but since the available space is fixed, what you really need is better organization. Add built-in cabinets and drawers. Hang jigs and infrequently used items high on the walls or even on the ceiling. Use the space between the joists or build low-profile racks for hanging things from the overhead. Use the floor underneath tools, the space beneath the staircase, and the area between the garage doors. Make the most of every inch of space, and your shop can happily coexist with the rest of your things.

Cabinets and shelves

Keep your Essential Shop space intact and use it to build cabinets to replace some of the shelving you're using. Build the cabinets in two banks and cover their tops with ¾" (or thicker) plywood or MDF so the tops are at bench height. Leave room between the banks for your bench, and you'll end up with a whole wall of workspace with surfaces at bench height. You can add more cabinets above the bench, but you'll have better light and more room to work if you don't. Use the

Think vertically. Use every bit of wall space—even the ceiling—and don't let the space beneath shelves or tools go to waste. The more you can store, the more room you'll have for woodworking.

space above the bench by installing narrow shelves or hanging frequently used tools there.

Rather than succumb to the desire to build the perfect storage cabinets with complicated features, start by building a batch of simple boxes with doors. Get them in service and figure out what wants to live where. To increase storage space later, you can customize the cabinet to its contents by installing partitions, sliding shelves, and drawers.

Now there's space on the other wall for storing lumber and sheet goods. Store

Floor Plan, the Basic Shop

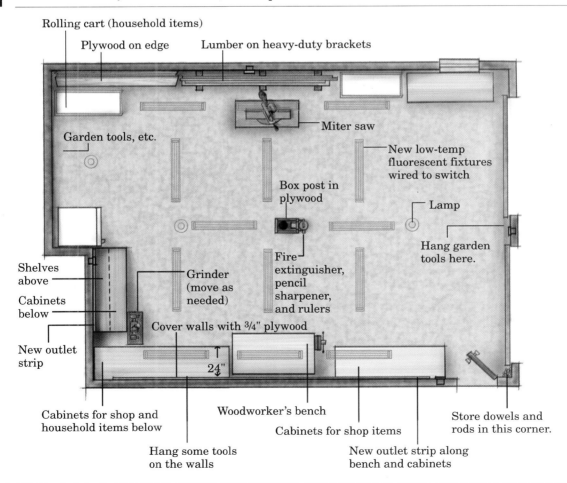

Rolling cart (household items)

Plywood on edge

Lumber on heavy-duty brackets

Garden tools, etc.

Miter saw

New low-temp fluorescent fixtures wired to switch

Box post in plywood

Lamp

Shelves above

Grinder (move as needed)

Cabinets below

Fire extinguisher, pencil sharpener, and rulers

Hang garden tools here.

New outlet strip

Cover walls with 3/4" plywood

24"

Cabinets for shop and household items below

Woodworker's bench

Cabinets for shop items

Store dowels and rods in this corner.

Hang some tools on the walls

New outlet strip along bench and cabinets

▲ **When your bench** is pushed against the wall with your most commonly used tools hanging nearby, you can get right to work whenever you have a spare moment. But sometimes you'll want to pull it away from the wall to get at all sides of your project.

▼ **Your first workbench** can stay in its original location, and you can reorganize the area around it for storing lumber. The sheet goods are stored in the far corner, accessible by moving the rolling shelves filled with home and garden items.

plywood at the back of the shop, leaning the sheets against the wall vertically—or as close to vertical as possible. Store solid wood on well-fastened heavy-duty shelf brackets. Twelve feet of shelving will accommodate lumber up to about 14' in length—longer pieces can rest on 2x4s set on the floor.

An array of clamp racks on the wall is an impressive sight, but wall space is at a premium in most shops. When you have only a few clamps, you can keep them in a bucket or bin and push them under the bench when not in use. But clamps in a bin end up in a tangled mess. A better solution is to build a clamp rack on casters. It frees up wall space and puts the clamps close by when you need them. When you don't, just push the rack aside.

Maximize the remaining space by parking some rolling shelves along the back of

ROLLING TOOL CABINET

Designed for auto shops, a multitiered metal tool cabinet works just as well in a woodworking shop. It keeps your tools nearby, even when you're not working at your bench. If you're concerned about unauthorized use of your tools, you can lock the cabinet up.

BINS AND CRATES

Moderate-size clear plastic bins with lids are a good way to store stuff in the shop. They're not big enough to get too heavy, and you can see what's in them. Open crates are smaller, more heavily built, and stackable, but small items will slip through the holes in the bottom and sides.

SOFT-SIDED BAGS AND TOOL ROLLS

Tools kept in a soft-sided bag aren't as likely to be damaged as those rattling around in a metal case. With pockets on the outside and divided space on the inside, a couple of moderately sized bags can hold most of your tools. Keep chisels and rasps in a tool roll with pockets.

DOWELS IN THE WALL

Rather than fiddling with pegboard and hangers that always fall out, put your tools on lengths of ½" dowel set into holes drilled in the wall. Don't expect this to work in drywall—cover it with a sheet of ¾" plywood first. Use nails or screws for hanging small items like rulers.

The bank of cabinets shown in this photo is nothing more than a line of identical plywood boxes screwed to the wall. A plywood top spans the bank and matches it to the bench height. Built quickly with a pocket-hole jig, these cabinets filled up almost immediately. MDF doors came later and were dirt-simple to construct—just cut the MDF to size and install the hardware.

Prebuilt cabinets are an even quicker way to add storage, and the cost can be even less than the simplest shop-built cabinets. You'll have to design your space to accommodate the cabinet sizes stocked by your local lumberyard or home center, but you can set up the shop in an afternoon.

Made from ¾" shop-grade plywood with MDF doors, these cabinets are serviceable and sturdy. You might say they have a rugged and functional handsomeness.

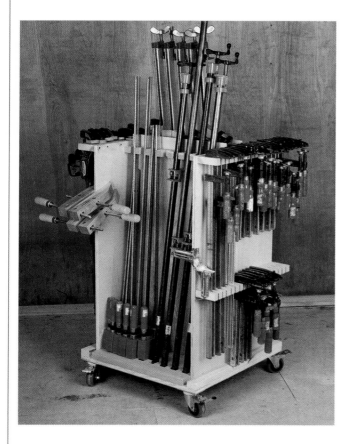

A mobile clamp rack saves wall space and puts the clamps where you need them: close by.

your shop with little or no space between them. When you need an item from a shelf, simply roll the cart out of line and into the open, much like opening a drawer.

Lighting

No matter how organized your shop is, you cannot do good work without adequate light. How much light is adequate? A lot more than you'd probably imagine.

Lighting design standards for a cabinet-making shop suggest that it have 1,000 lux of illuminance—in practical terms, it should be at least as bright as a supermarket. For comparison, the suggested illuminance for general office space or kitchens is 500 lux. Moonlight measures about 1 lux.

Recycle empty drywall buckets or buy new ones, but keep a few on hand. Though you can get nifty organizers for carrying tools in a bucket, I prefer to use them for other things.

■ Use them for storing small clamps—they're convenient for carrying them to the workbench.

■ With the appropriate trays, they're a great way to store fasteners.

■ Use them as a storage bath for waterstones.

■ Manage extension cords by threading the pronged end out of a hole near the bottom and coiling the cord in the bucket.

■ Store used solvents in them until your town has a hazardous-waste-disposal day.

■ Filled with mineral spirits and kerosene, they can keep your best paintbrushes clean and ready to use.

■ Filled with sand (or water) they serve as weights for clamping or veneering.

■ Turned upside down, they make great seats or makeshift sawhorses.

How do you translate suggested lux into the number of fixtures you need in your shop? A lighting designer would use the lux number and work through several equations, taking into account the size and construction of the room, the wall and floor color, the fixture's design, and other factors, and come up with a shopping list.

You can boil it down to this: Get one 4' double-tube fluorescent fixture for every 36 sq. ft. of shop space. If your shop walls are dark, or if the overhead is open-joist work with no ceiling, you'll need 50% more fixtures. When you do the math, round up and err on the side of more fixtures because aging eyes need even more light. To see the fine marks on a scale, a 70-year old needs twice as much light as a 30-year old.

Position the light fixtures around your shop to avoid dark corners. Put a line of lights down each side of the shop to illuminate your benches and storage and hang the rest in the middle of the room. You may not end up with even spacing because of things like garage-door tracks, ductwork, or beams. Just work with what you have and position fixtures so the obstructions don't block too much light. Finally, use task lighting whenever you need a little more clarity.

The Efficient Shop

What to Consider

After spending time working in your Basic Shop, you're probably aware of its limitations. You've become an expert at ripping with a circular saw, but you dislike all the work needed to set up the sawhorses, extension cords, and foam for a few cuts. You look forward to stepping up to a tablesaw, locking the fence at the desired width, and ripping away without all that set-up time. You're eager to get more time on woodworking by reducing the tedium and focusing on the fun parts.

You've also built enough projects to begin to understand wood and its ways, and you probably want more control over your materials. You're ready to buy rough lumber and mill it to your own specifications so it's flat, straight, true, and dimensionally stable. Maybe you're even thinking about buying thick boards and resawing them into thinner pieces so all the wood in a tabletop or set of drawer fronts matches in shade, tone, and grain. If so, it's time to add some big machines to your shop.

Jointer and planer make a pair

Good work starts with clear, straight-grained wood that's machined flat, and to do that you need both a jointer and a planer. Jointers flatten, straighten, and square up an edge,

PROS AND CONS OF THE BANDSAW AND TABLESAW		
	BANDSAW	**TABLESAW**
PROS	No kickback Cuts curves Resaws Does not require a fence or guide for every cut Cuts angles safely and easily Is quiet Has small footprint	Rips cleanly (needs only a few strokes with a plane to smooth) Makes accurate crosscuts Cuts clean joints with the right jigs Cuts dadoes and grooves
CONS	Ripping cuts need to be planed flat and smooth after sawing Needs many (simple) adjustments Small table sizes make handling sheet goods difficult	Is prone to kickback Needs accurate adjustments for safety Requires a fence or guide for every cut

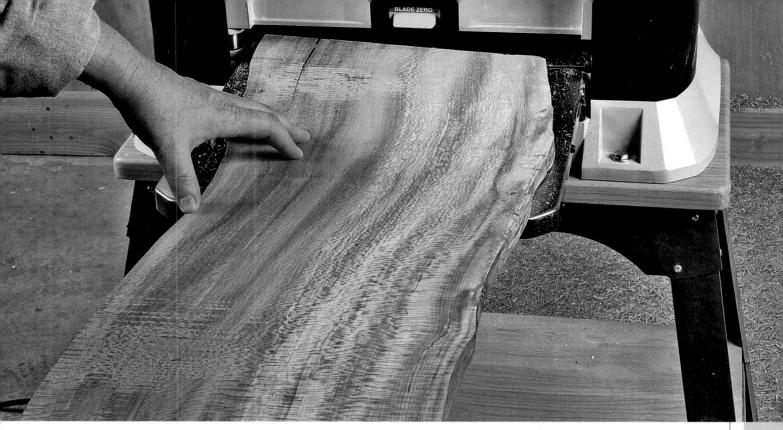

You can select lumber and harvest pieces to showcase the grain when you have the tools to dimension rough lumber to your own specs.

but they don't ensure that both faces are parallel to each other. That's why you need a planer. A planer simply renders the top of a board parallel to its bottom. In some cases it can remove twist or cup, but you can't count on it (see "Five Steps to Four-Square Lumber" on p. 95).

Tablesaw or bandsaw next?

After planing, a board is flat, uniformly thick, and has one good edge that's square to both faces. The next step is to run the good edge against a fence and rip the board to width on a tablesaw or bandsaw. Which should you have in your shop? Both do the job; the one you pick is more a matter of philosophy than fact.

Many novice woodworkers prefer the bandsaw because it's more forgiving: An error in guiding the work means a wavy saw kerf and more planing in the next step. A similar error on a tablesaw could result in a serious injury (see "Pros and Cons of the Bandsaw and Tablesaw" on the facing page).

I could make a strong argument in favor of either tool; in the end you'll choose the saw that suits your woodworking style. If you're partial to sculpted shapes, enjoy carving or turning, or plan to build a boat, you'll want to start with the curve-cutting bandsaw. If you're into the Arts and Crafts style, or want to build case goods, buy the tablesaw first.

Bottom line: Eventually you'll have both.

Add other tools as needed

After you've acquired the machinery for dimensioning lumber, put away your cash (or plastic) for a while. Plan to buy the other tools, but not until you need them. For instance, you can put off the drill press for a while, but when perpendicularity becomes a crucial attribute for the holes you need to drill, it's time to get one. The same is true for the router table and the mechanic's tools.

MACHINES USED TO MILL ROUGH LUMBER TO SQUARE

- Planer
- Tablesaw
- Bandsaw
- Jointer

MACHINES THAT CREATE JOINTS

- Tablesaw
- Drill press
- Router table
- Bandsaw
- Jointer

MACHINES THAT CREATE DECORATIVE EFFECTS

- Table-mounted router
- Tablesaw
- Drill press
- Bandsaw

A funny thing about getting a new tool—once you've learned to use it, you'll find it's indispensable. It becomes part of your problem-solving arsenal, and you can't imagine how you got along without it.

As your shop gets more efficient, it also gets more complex. Adding these tools means you'll need considerably more space, not only for the tools, but also for the larger projects you'll undertake because of them. You'll have to think about where you'll use the tools and how you'll store them. Electricity is suddenly an issue, and you may have to upgrade your wiring or add new circuits. You'll need better lighting, as well as enough heat and ventilation to keep you comfortable.

It will become an Efficient Shop, a place where work is streamlined to minimize the tedium and maximize the fun. But don't become so concerned with efficiency that you turn your woodworking into another source of pressure in your life. Woodworking shouldn't be about having the most powerful machines or how quickly you can complete your projects—those are the concerns of production engineers. Woodworking as a hobby is instead about slowing down, connecting with the wood, and challenging your hands, your head, and your heart.

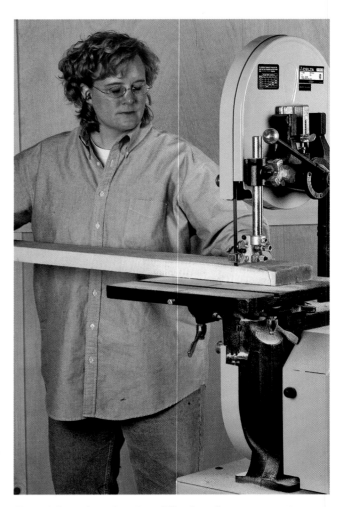

Resaw logs into lumber. The bandsaw can cut angles and curves, crosscut, rip, and resaw thick pieces of wood into thinner pieces.

FIVE STEPS TO FOUR-SQUARE LUMBER

FLATTEN FACE SIDE

Run one face over the jointer as many times as necessary until it's smooth. With proper jointer technique, that face will come out flat in length and flat in width with no warp, twist, or bow. Mark it as the face side.

FLATTEN FACE EDGE

Place the face side against the fence (make sure it's 90°) and run one edge through the jointer until it cuts along the full length and full width of the edge. Mark it as face edge.

PLANE TO THICKNESS

Place the face side down on the table when passing the wood through the planer. It shaves the top until it's parallel to the face side. Make the last pass on the face side, because the planer leaves a smoother cut than the jointer.

RIP TO WIDTH

Put the face edge against the tablesaw or bandsaw fence and rip to the desired width. Take a few swipes with a hand-plane to remove the tool marks and prepare the edge for gluing.

CUT TO LENGTH

Use a miter saw or a miter gauge on the tablesaw to make smooth, square cross-cuts to length.

Bandsaw

The bandsaw is the champion of cutting curves, but it can also rip and crosscut, as well as cut angles, compound shapes, and joints. It's also the best way to resaw a thick piece of wood into two thinner pieces. In short, it's one of the most versatile tools you can own. It can do nearly everything a tablesaw can do, except cut grooves. And the bandsaw is a friendly, relatively safe tool that doesn't take up much room in a shop.

Bandsaw basics

The "band" of the bandsaw is the flexible steel blade that wraps around two rubber-clad wheels. A motor turns the lower wheel, rotating so the blade moves downward at the point of cut, holding the workpiece on the table. Unlike a tablesaw, the bandsaw has no tendency to pick up the workpiece and fling it across the shop.

Guides keep the blade from wandering, as does tensioning the blade. This is done by tightening a screw to raise the top wheel.

WATCH OUT

- Hook your shop vacuum up to the dust port during ripping and resawing.
- Watch the position of your fingers while cutting angles or tapers.
- Round stock can rotate, pinching your fingers.
- Cutting too tight a curve will cause the blade to twist in the guide, bind in the wood, and possibly break.
- A ticking noise usually means the blade will soon break.
- Overtensioning will wear out the saw's bearings prematurely.

This puts enormous stress on the saw's frame, and it must be strong enough to hold this tension. The traditional material for bandsaw frames is cast iron, a heavy material that also helps to dampen the vibration caused by all the rotating parts. These days

WHAT A BANDSAW CAN DO

■ CUT CURVES

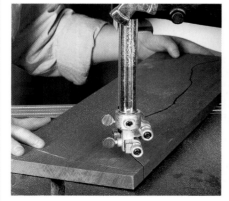

Cut open-ended curves and scroll.

■ RIP

Cut along the length of a board.

■ RESAW

Cut thick pieces to thin.

the trend is toward lighter-weight frames that get their strength from welded web frames.

When you refer to the size of a bandsaw in inches, you're talking about three measurements. A 14" bandsaw has 14"-diameter wheels, its throat width allows just less than 14" for a cut between the frame and the blade, and the table is about 14" square.

What to buy

Buy a 14" saw and you won't need to trade up unless you want to resaw boards wider than 12". Don't bother with smaller saws; they just don't have the capacity for ripping or resawing. A larger saw is always nice, but there's a big price jump up to a 16" or larger saw.

The standard 14" bandsaw can resaw boards up to about 6" wide. An optional riser block bolts between the upper and lower castings to increase the cutting height to almost 12". You'll want it sooner or later, so just get it when you buy the saw.

A 1-hp motor is standard on these saws; however, it can stall when cutting thick wood or resawing. Go for a 1½ hp from the start and save the trouble and expense of changing it later.

Prize lumber from firewood. Most woodpiles contain some treasures—sweet-smelling fruitwood or beautiful burled or spalted pieces. Screw the log to a sturdy right angle jig and resaw it on your bandsaw.

Every bandsaw has two sets of guides that support the blade both above and below the cut (one set is under the table). Look for guides that adjust easily, because you'll need to reset the guides whenever you change blades.

■ **CUT ANGLES**

Cut angles and compound curves.

■ **JOINERY**

Cut joints like this tenon freehand or with jigs.

The 14" Bandsaw

The 14" cast-iron bandsaw is the workhorse machine in small shops worldwide, both amateur and pro.

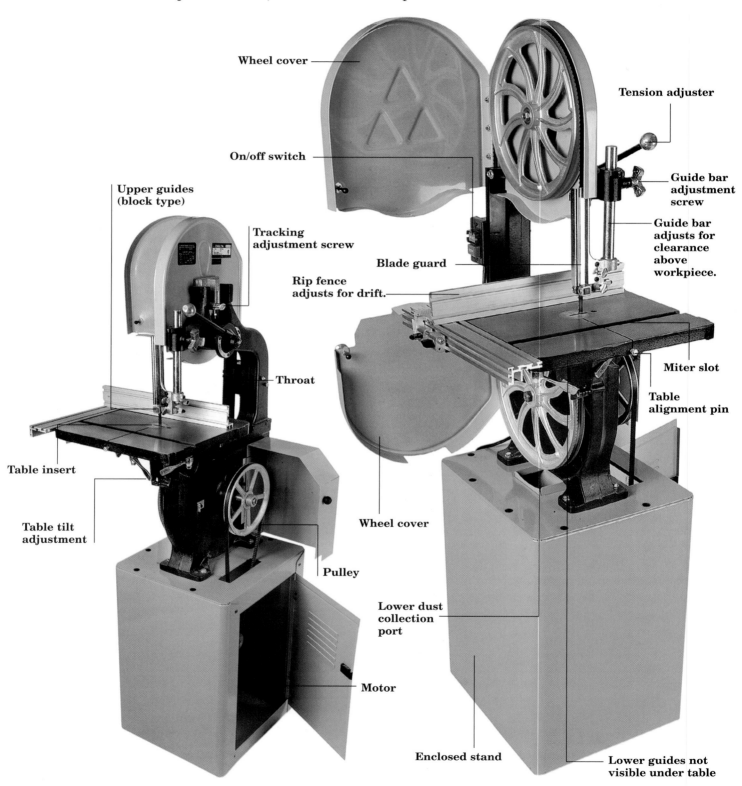

Wheel cover

On/off switch

Upper guides (block type)

Tracking adjustment screw

Blade guard

Rip fence adjusts for drift.

Throat

Table insert

Table tilt adjustment

Wheel cover

Pulley

Lower dust collection port

Motor

Enclosed stand

Tension adjuster

Guide bar adjustment screw

Guide bar adjusts for clearance above workpiece.

Miter slot

Table alignment pin

Lower guides not visible under table

BALANCED WHEELS AND PULLEYS

Just like the wheels of a racing bike, the wheels and pulleys on your bandsaw should be well made and balanced for smooth rotation. Look for machined surfaces and feel the back of the wheels for spots that have been drilled out to balance the wheel.

BLADE TENSION

A sloppy blade will flutter and bow as it cuts. For best results, it must be adequately tensioned. As a rule of thumb for home-shop-sized saws, tension your blades to the second-highest mark on the tensioning scale, and don't use a blade wider than ½".

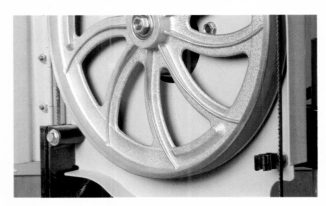

BLADE TRACKING

Once the upper wheel is tensioned, adjust its tilt to keep the band centered on the tire. Rotate the wheel by hand and turn the tracking adjustment screw (to the left of the spring tension scale in the photo directly above) until the band settles in the middle of the tire. The tracking knob (to the left of the spring in the photo directly above) tilts the upper wheel so the band runs around the middle of the wheels.

GUIDES

Limiting side-to-side and backward motion keeps the blade aligned and the cut precise. Adjust the side guides so they're close to the blade but not touching it. Guides take a lot of abuse and need frequent attention to do their job right.

Blades for the Bandsaw

The right blade is crucial for bandsaw performance. The wrong blade will cut slowly, with lots of heat and dust, and it won't cut the curve you want. You'll end up with a variety of blades to suit the many jobs a bandsaw can do, changing them as the need arises. Here's what you need to know to find the right ones.

Select the right pitch Pitch, the number of teeth per inch (tpi), is crucial to getting a smooth cut. The blade should have between 6 and 10 teeth in the work. For cutting a ½"-thick board, you'll use a blade with 12 to 24 tpi. If you're cutting a 2"-thick piece, use 3 to 5 tpi.

Tooth form Bandsaw blades come in different tooth shapes. Regular tooth blades have a straight cutting edge and deep gullets. They're good for general duty. A skip tooth blade also has a straight cutting edge, but there's a sharp angle between the tooth and the gullet. This makes for faster clearing of chips, so it's a good resaw blade. The hook tooth blade has wider spaced teeth and the cutting edge is undercut. The front of the tooth swoops into the gullet in a sharp curve, making for an aggressive cut suitable for very hard woods.

A typical blade. **This ½" x 3 tpi bimetal hook-tooth blade is good for both resawing and sawing thick stock.**

Width The wider the blade, the straighter it cuts. For the most part, stick with blades ½" wide or narrower. Wider blades require more force to tension than a small-shop bandsaw can generate. For cutting curves, you need a width suited to the desired radius. A ½" blade can cut a 2½" radius curve; a ¼" blade cuts a ⅜" radius.

Material Carbon steel is the most common and least expensive material for bandsaw blades. It's fine for general-purpose sawing, but it won't last long for resawing.

Bimetal and hardened blades use softer steel for the majority of the blade, with a band of harder steel at the teeth. They cost more and stay sharp longer.

To get the most benefit from the bandsaw's flexibility, you'll need to change blades often. It's not difficult with practice—the key is to do it in an orderly way. First things first: Unplug the saw.

PRELIMINARIES

First, release the tension, using the wheel or lever. Move the blade guides and thrust bearings out of the way. Then remove the throat plate and the level pin at the end of the table slot. Finally, ease the blade off the wheels and thread it through the table slot.

This is a good opportunity to do some basic maintenance. Clean dust and pitch from the tires. Also, clean the guides and make sure they're in good working order.

MOUNTING THE BLADE

While it may seem obvious, make sure you install the blade in the right direction. The teeth should face down toward the table. If they don't, the blade is inside out. Tension the blade and rotate the upper wheel by hand to test that the blade runs on the center of the tire. Adjust the tracking as necessary. Replace the throat plate.

ADJUSTMENTS

Adjust the upper and lower guides and thrust bearings. The thickness of a dollar bill sets the distance between the guide and the blade. The thrust bearing at the back of the blade should just kiss the back of the blade before cutting. It's wise to check the table for square rather than rely on the built-in protractor gauge.

If you use a rip fence, you'll need to adjust its angle to match the blade's cut. Choose a straight-edged piece of wood and draw a line parallel to the edge. Saw about halfway down the line (a little waviness is okay—the general trend is what matters) and stop the saw while keeping the board in place. Adjust the fence so it's against the board (see the photo below right). Finally, clean and wax the table.

Set the guides. A folded dollar bill is an inexpensive alternative to a feeler gauge for setting the correct distance between guides and the blade.

Reset the rip fence. Each time you change the blade, you'll have to adjust your rip fence to match the blade's cut.

Tablesaw

The tablesaw is the center of many woodworking shops. You may already know that it's the most efficient tool for ripping wood to width, but it does a lot more than that. Its ability to saw straight lines makes it a great way to cut grooves along a board's length, or dadoes across its width. If you need a big rabbet, a tablesaw can do it in two saw cuts. With a good crosscut sled running in the miter slot, the tablesaw is perhaps the most accurate way to crosscut or miter. When the face side of a board is run at an angle over the blade, it can cut an almost infinite variety of cove moldings. Add the ability to cut bevels, and you get a tool that not only cuts lumber to size, but also cuts complex joints and molds edges. No wonder it's so widely used.

But the tablesaw has a dark side—it does not tolerate errors. Ignoring proper technique or failing to attend to details can result in serious injury. Many beginning woodworkers fear the tablesaw with good reason, but acceptance is a better attitude. The turning blade's power is inexorable; you can't stop or change it, but you can harness it. If you understand the tool and use it on its own terms, you'll enjoy a lifetime of safe and successful sawing.

What to buy

Get a 1½-hp contractor's saw with a mobile base, all the optional cast-iron tables, and the largest side extension table you can fit into your shop. You'll never regret buying capacity. A few brands offer optional cast-iron sliding tables, ideal for crosscutting; just keep in mind that they take up a lot more room.

For about half the money, you can get a 10" portable benchtop saw. They're at their best with surfaced lumber that's less than about 1" thick, and they will need babying to

WHAT A TABLESAW CAN DO

■ RIP

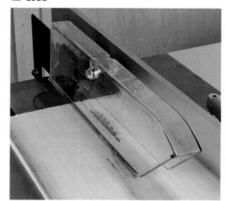

Run a jointed edge against the fence, and rip parallel to that edge.

■ CROSSCUT

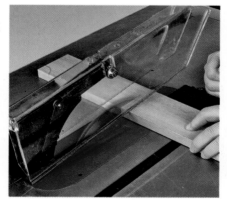

Cut across the width, always guiding the cut with a device running in the table's slot.

■ CUT ANGLES

The blade angles to 45°. Position the fence so the blade angles away from it.

handle more. The small direct-drive motors are loud, prone to overheating, and easy to bog down. A benchtop isn't a bad saw to start out with—the price is right and it's not too intimidating. But you'll soon outgrow it.

If you're flush with cash, consider getting a cabinet saw, the top-drawer choice. You'll

Stand to the left of the blade when ripping and use featherboards to hold the board against the fence. Feed the board steadily, pushing against the fence just behind the blade.

pay twice as much as you would for a contractor's saw, but it will hold its resale value. With 3+-hp motors driven by two or three

■ **CUT MITERS**

Using a miter gauge running in a slot, you can crosscut any angle.

■ **CUT GROOVES AND RABBETS**

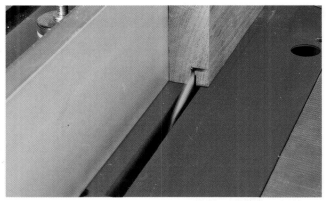

Cut grooves, dadoes, or rabbets.

Contractor's Tablesaw

A solid contractor's saw handles most of the tasks a woodworker will tackle, at about half the cost of a cabinet-style saw.

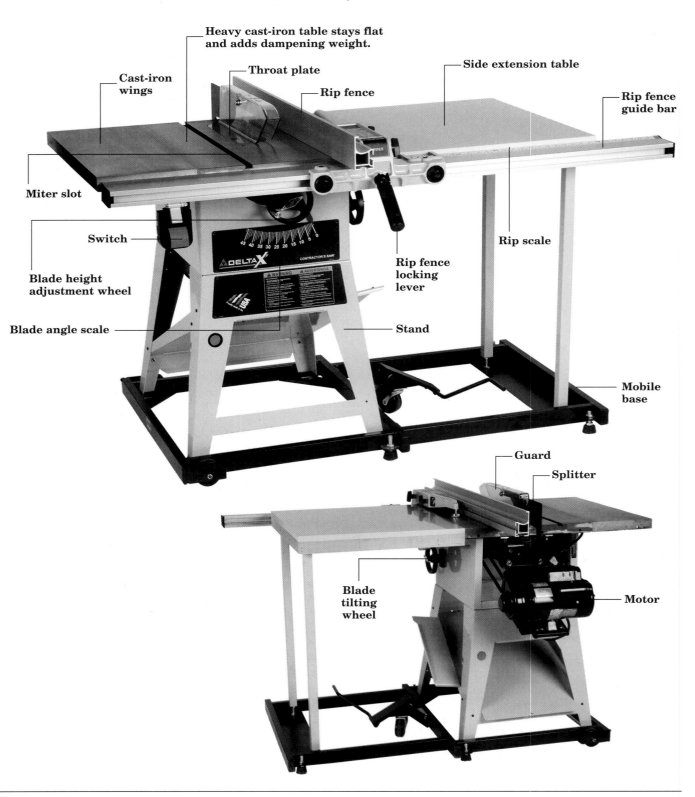

Heavy cast-iron table stays flat and adds dampening weight.

Cast-iron wings

Throat plate

Rip fence

Side extension table

Rip fence guide bar

Miter slot

Switch

Blade height adjustment wheel

Blade angle scale

Rip fence locking lever

Rip scale

Stand

Mobile base

Guard

Splitter

Blade tilting wheel

Motor

Benchtop tablesaw. A 10" benchtop saw is capable of accurate but light-duty work. It's affordable, portable, and easy to stow under a bench or even on a shelf.

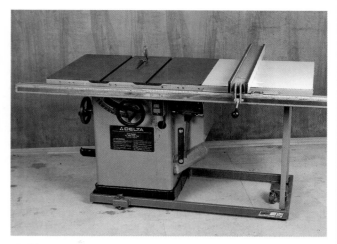

Cabinet-style tablesaw. A 10" cabinet saw is heavier, more accurate, more powerful, and easier to adjust. With a mobile base, you can easily move it around your shop.

belts, these saws have the guts to saw through anything and do it all day without overheating. Their greater mass dampens vibration, and the tilt and depth mechanisms (bolted to the cabinet) are engineered to be tough, accurate, and easy to adjust.

On contractor's and benchtop saws, the depth and tilt mechanisms are bolted beneath the table. They're notably fussy, and it can be difficult to get and keep the blade parallel to the miter slot, especially after tilting the blade. Work around this by locking the blade in the vertical position and building sleds for cutting angles, as shown in the second photo from the top on p. 107.

No matter what saw you choose, get the fence upgrade (see "Tablesaw Accessories" on pp. 146–149), or buy an aftermarket fence. Include dust collection ports or skirts as appropriate, and get at least two blades—a rip blade and a 40-tooth combination blade.

Safety

The tablesaw presents two distinct hazards: the exposed blade and kickback. The exposed-blade hazard is simple to avoid: Keep the blade covered and keep your hands away

from it. Every new tablesaw comes with a blade guard; use it. If you find it clumsy or poorly designed, replace it with a better one (see "Tablesaw Accessories" on pp. 146–149). Keep your hands at least 6" away from the blade and use push sticks and featherboards to hold and control the work safely.

Kickback is a more complex problem. It happens when the wood binds and catches on the blade's back edge. The rotating blade first lifts the wood up and, as the rotation

THE FIVE LAWS OF THE TABLESAW

- Every rip requires a fence, and only straight edges go against the fence.
- Every crosscut involves the miter slot in some way—don't use a fence or cut freehand.
- Always guard the blade; use a splitter when possible.
- Keep your hands away from the blade and use push sticks, hold-downs, and featherboards.
- Don't use the tablesaw when sickness, medications, anxiety, or fatigue might impair your mental agility.

Featherboards and push sticks **guide the board and keep hands from getting too close to the blade. At the far right is a shopmade push stick. To its left is a plastic push stick. Aligned in the slot are a selection of featherboards: a wooden single-lock featherboard, a more secure double-lock variety, a shopmade featherboard that clamps to the table, and a magnetic featherboard.**

continues, flings it back toward the operator. Sometimes the trajectory is low and the piece hits at hip level; sometimes it's at chest or head level. If you're standing in the right place (to the left of the blade), it'll pass you by, but others may not be so lucky. A kicked-back board moves at around 120 mph—more than enough to seriously injure someone on the other side of the room. Perhaps worst of all, if your hand is too close when the kickback starts, the initial upward motion can draw it into the blade. Kickback is scary, but you can prevent it by using a splitter, setting up your saw properly and using proper sawing techniques (see the top photo on p. 103).

Ripping

- Use a splitter and guard
- Run a jointed edge against the saw
- Use a featherboard to hold the wood against the fence right at the blade

- When sawing, push stock toward the back edge of the fence
- Lean against the saw for stability
- Don't reach; walk around the saw to collect your pieces

Crosscutting

- Use a miter gauge, crosscut sled, or crosscut box
- Never cut freehand
- Never crosscut against the fence
- Use clamps to hold your work against your sled or jig
- Use a guard to reduce blade exposure

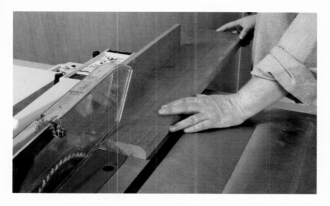

RIP FENCE

The tablesaw cuts parallel to the fence, so be sure to run a straight, jointed edge against it whenever ripping stock. A good fence should be so accurate you don't need to measure each cut but can simply read the measurement off the scale.

ANGLE SLED

Cutting bevels can be tricky, but a sled like this one makes it easy. By keeping the blade at 90°, the torsion bars in the undercarriage won't rack and cause misalignment. It also ensures accuracy and is safer.

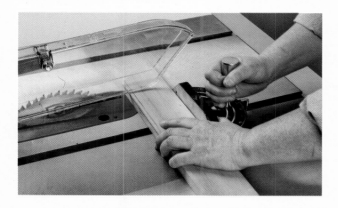

MITER GAUGE

The miter gauge slides in the slot and rotates to cut angles. For best results, attach a sliding auxiliary fence and adjust it close to but not touching the blade.

CROSSCUT SLED

The safest and most accurate way to crosscut is to place stock on a sled that rides on runners in the miter slots. A well-equipped shop accumulates several sleds—for small pieces, for 45° miters, for dados, for long pieces, for wide pieces, and so on. See more on crosscut sleds in "The Crosscut Sled" on p. 148.

Jointer

While it can bevel, taper, and even rabbet, the jointer's most important job is flattening the faces and edges of boards. A well-built, well-adjusted jointer removes twist, cup, bow, and crook during the crucial first two steps of properly milling lumber (see "Five Steps to Four-Square Lumber" on p. 95).

What to buy

Jointers are precision tools and fussy to maintain. If the relationships between the tables, knives, and fence are not all perfect, the tool can't produce a flat surface. So your first concern when buying a jointer should be that the fit and finish are good enough to allow the necessary fine adjustments. Make sure the table and fence are flat—check them with a metal straightedge and reject a tool that can't pass this test. Smooth mating surfaces where the tables slide is crucial for fine

adjustments, and the fence must move freely across the table's width.

A long table handles lengthy boards with ease, and three knives cut more smoothly than two. Don't even consider a jointer with no outfeed table adjustments—it will be much harder to get and keep the proper adjustments so critical to proper operation. A cutterhead lock is a nice feature—it pins the knives at top dead center for easier adjustments.

WHAT A JOINTER CAN DO

■ FLATTEN THE FACE SIDE

Removes cup, twist, and bow from the face side.

■ FLATTEN/SQUARE AN EDGE

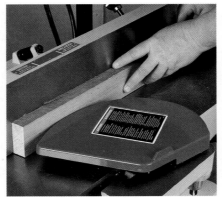

Removes bow and twist from the face edge and squares it to face side.

■ BEVEL AN EDGE

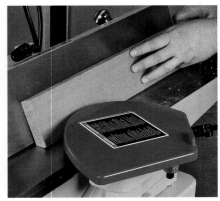

Angle the fence to bevel the edge of a board.

The 6" Jointer

Because it's the first step in milling four-square lumber, your jointer's size determines the maximum width board you can properly dimension.

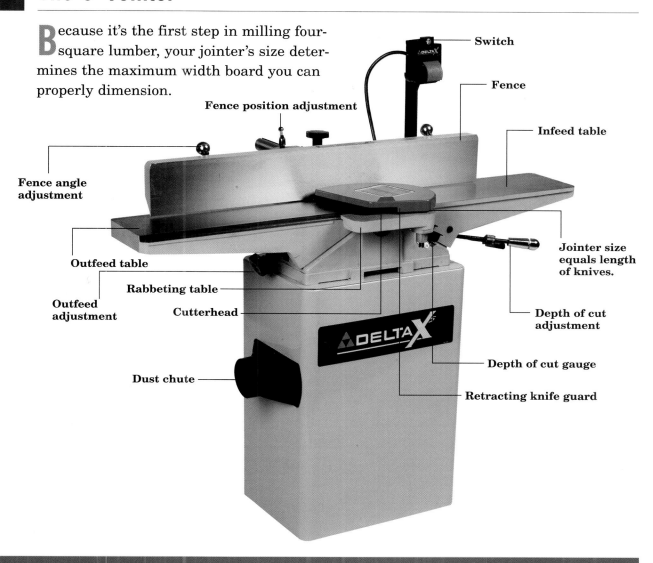

Switch

Fence

Fence position adjustment

Infeed table

Fence angle adjustment

Jointer size equals length of knives.

Outfeed table

Rabbeting table

Outfeed adjustment

Cutterhead

Depth of cut adjustment

Depth of cut gauge

Dust chute

Retracting knife guard

■ TAPER

A clamp defines the starting point to taper a long piece.

■ RABBET AN EDGE

Remove the knife guard to cut shallow rabbets.

Planer

A lot of people confuse the tasks done by a jointer with those done by a planer. A planer reduces a board to uniform thickness, with the top parallel to the bottom. It doesn't make the board straight by removing bow or twist—only a jointer can do that. To get a board both flat and square, you'll need both a jointer and a planer (see "Five Steps to Four-Square Lumber" on p. 95).

The planer is a simple machine and easy to use. As powered rollers pull a board over the bed of the planer, a spinning cutter-head removes material from the top of the workpiece. With proper knife and roller adjustment, you'll end up with a board that's uniform thickness and smooth from end to end.

What to buy

Get a portable planer that can handle boards 12" to 13" wide. For less vibration and smoother cuts, buy a machine with a cutter-head lock, and be sure to use it.

The number of times a knife cuts the wood also affects smoothness. The more cuts per inch (cpi), the smoother the surface. A low cpi means that the board moves quickly—a setting that's good for rough surfacing. For a finer finish, increase the cpi by slowing down the feed rate.

Sharp knives are the most important factor in smoothness. Buy a planer that makes changing blades as easy as possible, and change them often.

To keep your shop clean, get the optional dust collection hood.

WHAT A PLANER CAN DO

■ FLATTEN THE TOP

With the jointed face down, plane the top smooth.

■ REDUCE WIDTH

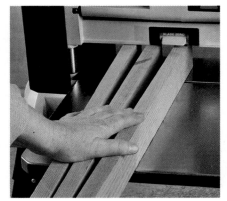

Reduce a number of boards to equal width.

■ REGULATE THICKNESS

Plane a board to uniform thickness.

The Portable Planer

Once one side of a board has been jointed, the planer cuts a flat smooth surface on the other side. It also ensures that the board has uniform thickness throughout its length.

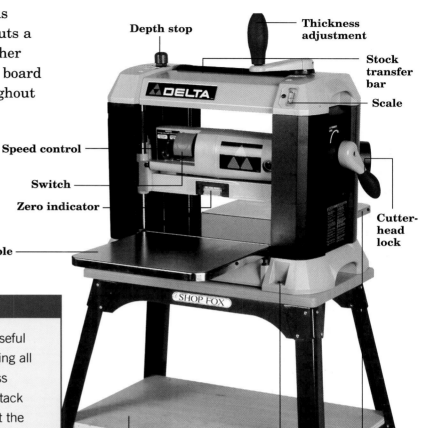

Depth stop

Thickness adjustment

Stock transfer bar

Scale

Speed control

Switch

Zero indicator

Infeed table

Cutter-head lock

Mounting holes to prevent tipping

Tool storage

Outfeed table

ALL TOGETHER NOW

Zero indicators and depth stops are useful features, but you can't count on having all the wood for a project the same thickness unless it's planed in the same session. Stack jointed boards on the infeed side and set the planer to take a light cut on the thickest board. Run every piece through, face side down, and stack them on the outfeed side. Carry the pile back to the infeed side, lower the cutter, and repeat as necessary. On the final cut, run the face side up to clean tool marks left by the jointer.

Planing it to size. The only way to ensure equal thickness is to plane all your stock at once.

Router Table

The router table is to the router as the tablesaw is to the circular saw. In both cases, mounting the tool upside down in a table and guiding the work over a protruding cutter produces something more versatile and functional than the original.

Most routing activities are easier to do on the router table (especially edge treatments), and you can rout pieces that are too small to do with the router in hand. A router mounted in the table can safely run big cutters with diameters too large to control by hand. A router mounted in a table also leaves smoother surfaces and cleaner profiles than one used on the workpiece. And if all that isn't enough, using the router table is faster because you're spared all that clamping and unclamping.

What to buy

Your first router table should be a simple benchtop affair. It's small, easy to assemble,

and can be stored under a bench when not in use. Just be sure to clamp it down to a stable surface when routing—otherwise, it might tip over. Look for one with a smooth, flat top that won't deflect under load and adjustments for leveling the insert plate with the top. Some router tables have blank inserts—you drill the mounting holes to fit your router. It's not a difficult process, but you save some trouble if you can get an insert that fits your router model.

WHAT A ROUTER TABLE CAN DO

■ MOLDINGS

Mold the edge of a board against the fence.

■ TEMPLATE CUTTING

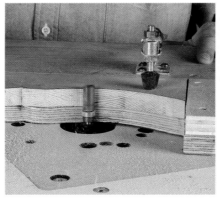

Cut to a template using a bearing-guided router bit.

■ RAISED PANELS

A vertical bit creates the raised field of a raised panel.

The Benchtop Router Table

The router table is little more than a worksurface that secures an inverted router on the underside of the tabletop. Add a fence and miter slots, and you've got a flexible tool that will get constant use in the shop.

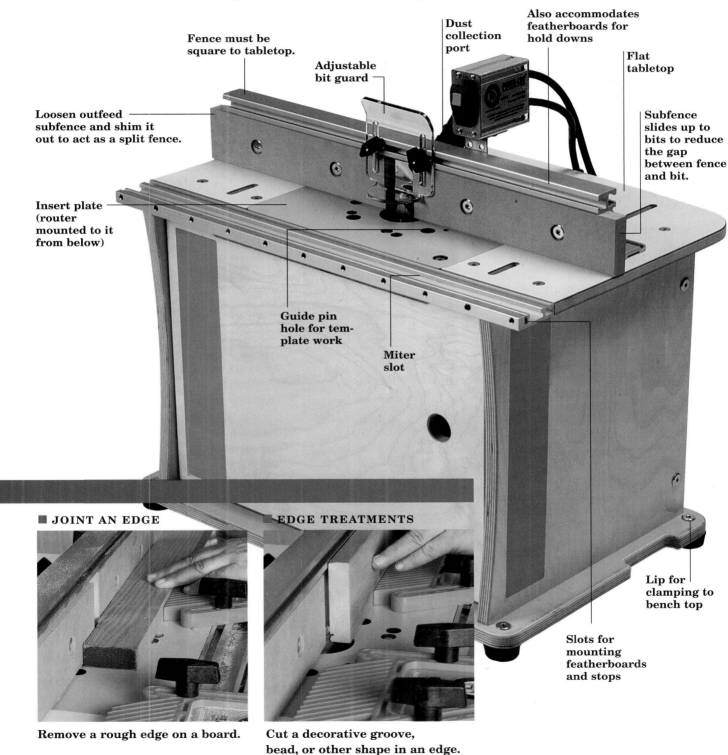

Fence must be square to tabletop.

Loosen outfeed subfence and shim it out to act as a split fence.

Adjustable bit guard

Dust collection port

Also accommodates featherboards for hold downs

Flat tabletop

Subfence slides up to bits to reduce the gap between fence and bit.

Insert plate (router mounted to it from below)

Guide pin hole for template work

Miter slot

Lip for clamping to bench top

Slots for mounting featherboards and stops

■ **JOINT AN EDGE**

EDGE TREATMENTS

Remove a rough edge on a board.

Cut a decorative groove, bead, or other shape in an edge.

The fence is perhaps the most important part of a router table. Make sure yours is straight and square to the tabletop, and that it is easy to adjust back and forth on the table. A replaceable wooden subfence that opens and closes to accommodate various diameter bits is crucial. It's also good for the fence to have T-slots for fastening feather-boards and stops, which not only make your work safer, but more accurate as well.

You can use the router you already have in your router table. To mount it, simply switch the plastic baseplate with the router table insert. It's a good idea to get an extra base to leave mounted to the insert—that way switching to the router table is simply a matter of slipping the motor from one base into the other.

Set the depth of cut. Adjust the bit height by raising or lowering the router in its mounting on the underside of the table.

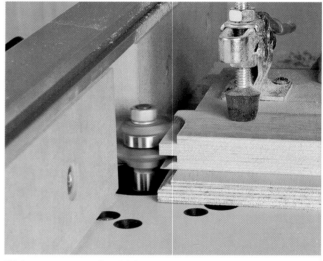

Using two matched bits **and two different setups, you can make complex joints that fit perfectly.**

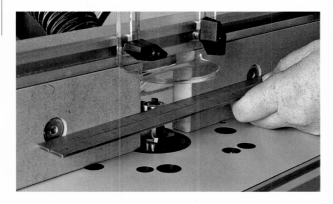

SOLID FENCE

When the molding process will leave some of the original edge on a board, a solid fence works best. This setup is best for rabbeting, molding, panel-raising, and cutting grooves. Here, a split fence is set up as a solid fence by aligning the two halves.

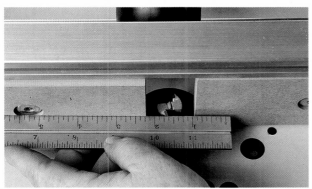

SPLIT FENCE

On a split fence, the infeed and outfeed sides of the fence are (or can be) offset. It's useful when you rout away all of an edge—for edge jointing or some profiling operations. You can build a split fence from two pieces of wood or make its functional equivalent by shimming out the auxiliary fence on the outfeed side of a solid fence.

GUIDE PIN

Remove the fence and use a bit with a pilot bearing, but don't start your cut freehand. Insert a guide pin near the bit and pivot the workpiece on the pin until it engages the bearing. Then you can ignore the pin and rout with the bearing.

MITER SLOT AND SLED

When you're dealing with square pieces, small pieces, or end grain, use a sled riding in the miter slot for control.

Drill Press

A drill press could claim a place in your shop if it merely drilled perpendicular holes. But it does so much more. It drills angled holes, runs a sanding drum or mortising attachment, and (unplugged) acts as a press for installing threaded inserts or tapping a hole. You'll use your drill press far more than you imagine.

Control your workpieces at all times by clamping them in place or restraining them against a fence. Otherwise, they'll rise with the bit and spin dangerously above the table. A fence is also useful for setting up indexes and stops for repetitive drilling.

What to buy

Bench space is at a premium in most shops, so get a floor-mounted variable-speed drill press with a 16" to 17" throat. It has the power and versatility to handle just about any task you'll encounter. Get one with a ½" chuck to handle large bits.

Make sure the cast-iron table has a hole in the center for the drill to pass through and slots for clamping. In addition to whatever table is standard, you'll want to add a wooden table with a fence and good clamping arrangements.

A quill lock is a huge time-saver. With it you can position the workpiece and pin it down by locking the bit in the full down position. This frees your hands to set stops, clamps, or indexing devices.

WHAT A DRILL PRESS CAN DO

■ DRILL PERPENDICULARLY

Get better results than you ever will by hand.

■ DRILL ANGLES

Tilt the table and lock it in place.

■ TURN A SANDING DRUM

Sand curves and complex shapes.

16½" Floor-Mounted Drill Press

Whether you're drilling a quick perpendicular hole or clamping jigs to the table to make complicated joinery on furniture parts, a drill press gets steady use in the workshop.

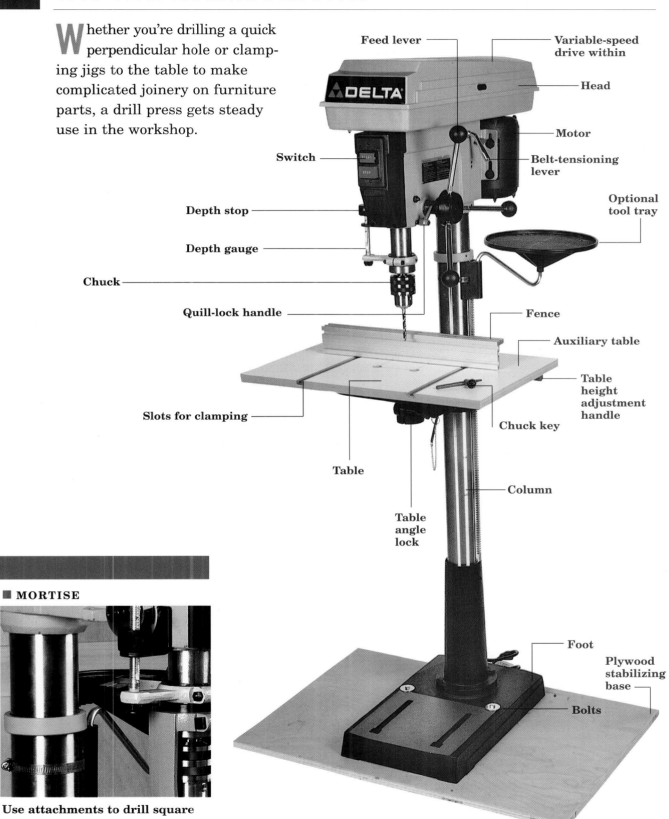

Feed lever

Variable-speed drive within

Head

Motor

Switch

Belt-tensioning lever

Optional tool tray

Depth stop

Depth gauge

Chuck

Quill-lock handle

Fence

Auxiliary table

Table height adjustment handle

Slots for clamping

Chuck key

Table

Column

Table angle lock

Foot

Plywood stabilizing base

Bolts

■ **MORTISE**

Use attachments to drill square holes.

Specialty Hand Tools

By now your collection of hand tools is solid; it can handle most situations. The tools featured here are more specialized. You won't use them often, but in some cases one of these tools may be the only way to do what needs to be done. When you need it, you'll thank the day you bought it.

What to buy

Keep a flush-cutting saw with fine teeth on hand for trimming pins, plugs, and overhangs of all kinds without marring the surrounding wood.

Your next plane purchase should be a shoulder plane. It's perfect for fitting tenons and cutting rabbets. The sides are machined flat and perpendicular to the sole, so you can get into tight corners by using it either upright or on its side. Some shoulder planes have a screw in the nose that releases the front of the plane to turn the tool into a chisel plane for cleaning up corners.

The spokeshave takes some practice to master, but it's the best tool for smoothing and shaping curved pieces. It can also be used like a plane for small jobs with no curves.

Sometimes a machine-made dado or groove is a little too narrow, and setting up the machine to remove a small amount of wood may not be feasible or efficient. You can easily trim the sides of the dado with a side-rabbet plane.

To round out your chisel collection, you should add a narrow ¼" skew chisel for working into corners and a ½" or ¾" cranked-neck chisel for getting into stopped dadoes and other tight places.

WHAT THESE TOOLS CAN DO

■ SAW FLUSH

A flexible saw with no set to the teeth trims flush without marring the surrounding wood.

■ FIT JOINTS

The blade on a shoulder plane cuts right out to the edge of the plane—ideal for fine-tuning joinery.

■ SMOOTH CURVES

A spokeshave is like a plane with a very short sole, so it can smooth curves a plane can't.

Specialty Hand Tools

Certain tasks have tools designed expressly for them. These tools might not be ones you'll use every day, but they often prove themselves indispensable.

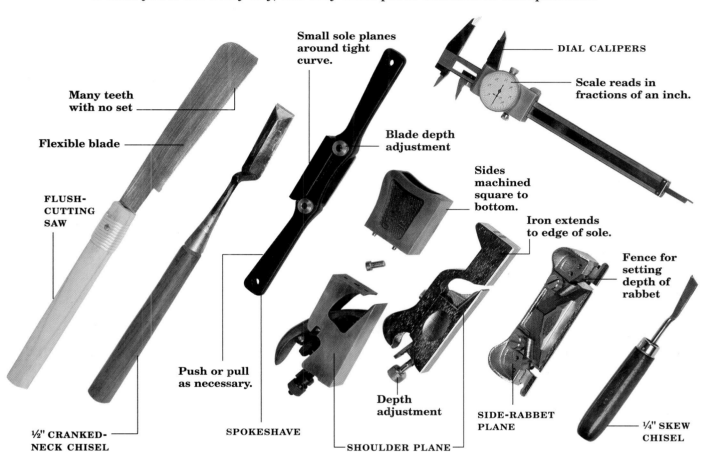

Many teeth with no set

Flexible blade

FLUSH-CUTTING SAW

½" CRANKED-NECK CHISEL

Push or pull as necessary.

SPOKESHAVE

Small sole planes around tight curve.

Blade depth adjustment

DIAL CALIPERS

Scale reads in fractions of an inch.

Sides machined square to bottom.

Iron extends to edge of sole.

Fence for setting depth of rabbet

Depth adjustment

SHOULDER PLANE

SIDE-RABBET PLANE

¼" SKEW CHISEL

■ TRIM DADOES AND GROOVES

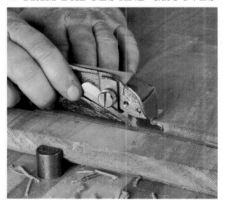

An edge-trimming plane is perfect for widening dadoes and grooves cut by machine.

■ PARE IN TIGHT CORNERS

Specialty chisels like the skew chisel or the cranked-neck chisel can get into tight places.

■ MEASURE THICKNESS

Dial calipers that read in fractions are the best way to measure thickness.

Mechanic's Tools

To maintain your woodworking tools, certain mechanic's tools are a must. Start your wrench collection by buying a set of combination wrenches with one open end and one box end, ranging in size from ¼" to about 1". Prices range from $20 to more than $600 for a set of 15. Your best bet is in the middle range, buying brand-name tools with a warranty.

Next, get a 10" adjustable wrench, a 10" Vise-Grip®, and some Allen wrenches. A set of ⅜" drive sockets lets you work in places where combination wrenches won't fit and is faster than wrenches for most situations. A ¼" drive set is also useful in tight spots.

Get a set of punches and drifts, or at least one of each (¼" diameter is a good size). Use the parallel-sided punches to pound stuck bolts out from the back of a workpiece.

A tapered drift is perfect for aligning bolt holes when setting up your machinery.

To keep tools well-tuned, you'll also need a dial indicator that reads in plus or minus thousandths on either side of zero along with a magnetic base. To simplify tablesaw alignment, you'll also want a base machined to fit into miter slots. Round out your mechanic's tools with a set of feeler gauges to measure small gaps, such as those between bandsaw blades and guides.

WHAT MECHANIC'S TOOLS CAN DO

■ **VISE-GRIP**

A versatile Vise-Grip works as a wrench, a clamp, or a vise.

■ **DRIFT**

Lever bolt holes into alignment with a tapered punch, also called a drift.

■ **FEELER GAUGES**

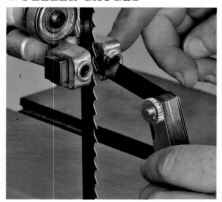

Part shim and part measuring device, feeler gauges measure small gaps.

Mechanic's Tools in the Woodworking Shop

Though they weren't originally designed with the woodworker in mind, certain mechanic's tools are a must for any woodworker.

⅜" drive socket set

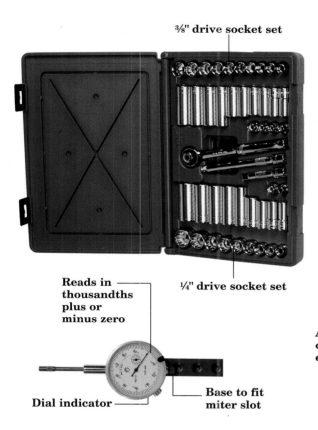

¼" drive socket set

10" adjustable wrench

6" adjustable wrench

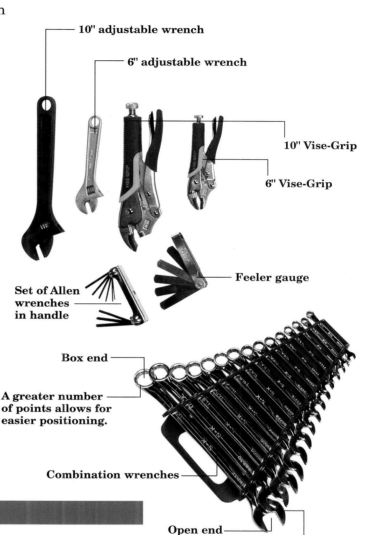

10" Vise-Grip

6" Vise-Grip

Feeler gauge

Set of Allen wrenches in handle

Reads in thousandths plus or minus zero

Dial indicator

Base to fit miter slot

Box end

A greater number of points allows for easier positioning.

Combination wrenches

Open end

Offset reduces skinned knuckles.

Set of individual Allen wrenches

■ ALLEN WRENCHES

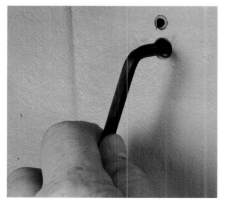

Use Allen wrenches to deal with the hex socket bolts on machinery.

■ WRENCHES WORK IN PAIRS

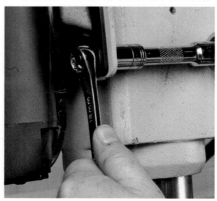

Use one wrench to hold the bolt, another to turn the nut.

The Efficient Shop Space

Adding all these machines to your shop complicates the space problem considerably. Once you move them in, the shop seems a lot smaller. And you can't just put your new machines anywhere. The tools have working relationships with one another, and they need to be located so that they can work together for maximum efficiency. If you're sharing your shop with cars, you'll have to figure out two locations for each tool—one for use and one for storage.

More machines means more load on your shop's electrical system, and you'll need circuits sufficient in number and size to handle the loads you'll put on them. Moisture is always a problem in home shops—too much or too little—and you'll have to monitor your shop and keep the levels in the proper range.

Finally, since you're spending more time in the shop, you need to think about how to keep it warm enough for your own comfort and safety.

Planning for flow

A look at any coffee-table shop book proves that there's no one way to set up a shop. Organizing a shop is an ongoing process; as you gain experience and develop your own woodworking style, your shop will change until the space works best for you. For now, you can place your machines by thinking about the steps involved in processing rough lumber into workpieces.

The processing starts with long, heavy pieces of lumber coming off the storage rack

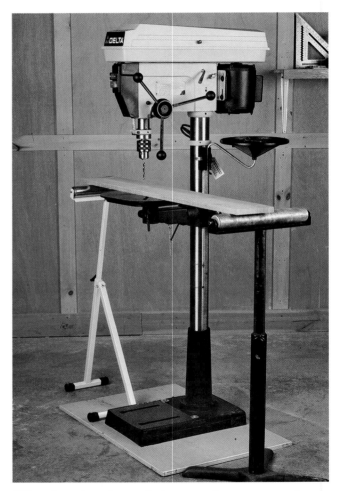

When locating the tools in your shop, you must consider the infeed and outfeed tables. Make sure there's enough room around a tool to manage a long workpiece, here supported by roller stands.

for rough cutting to length. To minimize carrying, place the miter saw nearby (and remember that you'll have to crosscut really rough lumber with a circular saw or a handsaw). The lumber then goes though the jointer for jointing the face side and then the face edge. Locate the jointer so you don't have to

Floor Plan, the Efficient Shop

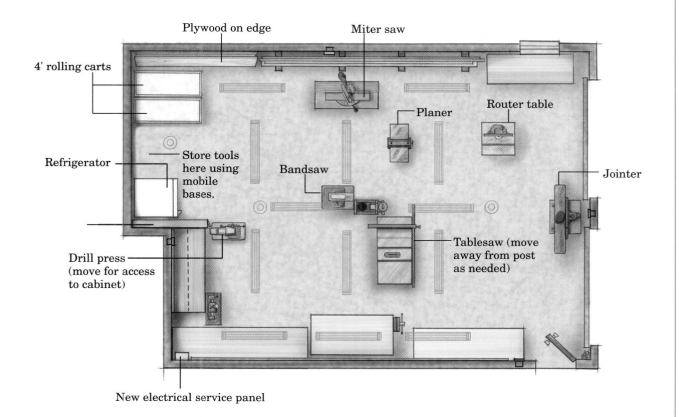

Plywood on edge

Miter saw

4' rolling carts

Router table

Planer

Refrigerator

Store tools here using mobile bases.

Bandsaw

Jointer

Drill press (move for access to cabinet)

Tablesaw (move away from post as needed)

New electrical service panel

When working on exceptionally long workpieces, you'll have plenty of room if you move outside.

Few small shops have enough room to allow the ideal infeed/outfeed area around each machine. With mobile bases, you can easily move the machines to accommodate a large workpiece.

negotiate corners or close spaces with a load of lumber, and keep enough space around it to stack the wood during both operations. After the jointer, move to the planer. From the planer outfeed, the stock goes to the tablesaw for ripping, and then to the bench or to an active storage location—depending on the scope of the project. Active storage could be on the shelf under the bench, on sawhorses nearby, or back in the lumber rack.

To plan the flow in your own shop, start by measuring your shop and drawing a rough floor plan to scale. Include doors, windows, cabinets, posts, and other permanent structures. Next, fashion rough cutouts of your machines' footprints (to scale), and play around with positions and locations. Don't forget that the amount of space you'll need to use the tool is greater than its footprint—

Concrete is not the best material for a shop floor. It's hard on the feet and legs—you'll ache after a long day's work. It's rough on any tool you accidentally drop, and it tends to hold moisture so you can't leave wood resting on it for very long. A wooden floor takes care of all those problems, plus you can drive screws into it and use it as a laminating surface or for tacking down supports that stabilize projects during assembly. Here's a quick and easy method of installing a wooden floor that won't reduce head-room by much and is easy to remove down the road if you move.

Lay a sheet of heavy plastic over a clean concrete floor, taping any joints with duct tape. Put 2x4s down on their wide sides and fasten them to the floor with a powder-actuated nailer or Tapcon® concrete screws. Build a grid on 12" centers and fasten ¾" plywood or underlayment, with the joints falling on the centers of the 2x4s. Then leave it alone, or paint it a light color.

A sheet of ¾" plywood makes an adequate ramp for the small difference in height between the shop floor and the driveway.

If your shop is in the garage and you want to roll tools outside on occasion, don't run the ply-wood up to the edge of the 2x4 on the driveway side. Leave a ledge to support a plywood ramp (see the photo above).

consider also the infeed and outfeed (see the photo on p. 122). Think in terms of handling sheets of plywood and solid wood up to 12' on a regular basis and larger workpieces on occasion.

Your floor plan won't show another important dimension—the heights of the tables—and this can be a major factor in machine placement. For instance, jointer tables are usually lower than tablesaw tables, so you can park yours in the tablesaw infeed/outfeed zone. You can even use it when sawing by fitting a shopmade bar of rollers to make up the difference in height.

In a small shop, a flow diagram might end up looking like a mad dance as the lumber moves though the milling process. That's

OK—you'll move a few steps, work a while, and move again. The point is to minimize carrying, especially around corners or in confined areas.

For more options, store your machines on mobile bases. Settle on a layout for everyday use and shift the tools for unusual situations. You can even move machines outside when necessary.

The shop in the floor plan drawing is the usual setup in my shop. It clusters the machines around the immovable post. To free up space on the bench side, the tablesaw overlaps the post. On the rare occasions when I need to cut at full width, I simply move the saw nearer the bench. The bandsaw situation is similar—99% of the time it works there;

if not, I move it. Located between the garage doors, the jointer has a little more than 9' of infeed and outfeed—plenty of room most of the time. The planer is the only machine that I regularly move—it stays against the wall until I need it. Though it's on a mobile base, the drill press rarely shifts location. On the base I keep a 5-gallon bucket filled with sand to lower the center of gravity so that the top-heavy tool is easy to move when necessary.

I like to keep the machines around the edge of the shop and leave the middle open for assembly (it's also easier to get the cars in). But if I'm working on smaller projects and the cars stay outside, I sometimes move the jointer and the planer to the center of the bay and nestle the back of the jointer to the right side of the planer. In that location, they're ready for immediate use.

Upgrading the circuits

Once you have an idea of where your tools will be used, you probably need to upgrade the electrical situation in the shop. After all, the space was not intended to be a shop. In most cases, you'll find that all the lights and outlets are on one 15-amp or 20-amp circuit shared with adjoining rooms. In that situation, running the bandsaw and the shop vac at the same time could trip the breaker and shut down not only the shop but also your kid's homework on the computer upstairs.

Lights, TVs, computers, and normal household items don't use much current and might never overload a circuit. Add a hard-working, power-hungry tool, and the combined current draw can easily exceed the capacity of the wire, causing it to heat and possibly burn (see "Current Draw for Common Woodworking Tools" on the facing page). The circuit breaker acts as a safety valve to spare the wire, shutting down when the current draw exceeds its limit. Simply installing a larger breaker is not a safe option; the wires are the weak link in the chain.

The best way to deal with this problem is to install a 100-amp line to the shop, with its own electrical panel and circuit breakers. While you're at it, add a 220v circuit for a heater and any larger tools you might add in the future. Run one 20A circuit down the bench side of the shop, another along the wall, with perhaps a third along the overhead and down the post. Install outlets every 10 feet or so and save yourself the clutter of extension cords. You can keep the existing lights and outlets on the household circuit, but label them as such and take care not to overload them.

Add a 100-amp subpanel in your shop and run your outlets on at least two circuits. To keep unauthorized users safe, you can turn off the circuits and lock the panel.

Every shop needs a humidity gauge (the one seen here is combined with a thermometer) to ensure that long-term humidity levels are moderate to keep your lumber from getting too wet or too dry.

With this setup, you can run more than one high-amp machine at once (planer and dust collector, for instance)—just plug them into separate circuits.

The moisture problem

In a perfect world, your shop would be 70°F with 40% humidity year-round. You'd be perfectly comfortable working, the wood in your lumber rack wouldn't absorb moisture or dry out, and your tools would never rust. But small shops are very much part of the real world. Because money is a serious consideration, we work in basements that are either too humid or parched by the household heating system, or in garages where the conditions are practically the same as the weather outside. We can't control shop climate, but we can improve it.

Moisture is your biggest concern, because it can cause the most damage to your tools (see the top photo on p. 128). In a small shop, rust forms during the spring and fall when humidity is high and temperatures change rapidly. Because of their mass of metal, tools change in temperature more slowly than the

One autumn in an unheated garage shop left a fine coating of rust on this jointer table and fence. Such rust is easily removed, but the table will never again be shiny.

air does. As on a glass of ice tea on a hot day, water droplets will form on the tools, and they will rust. The best way to prevent such rust is to keep the tools warm. I've found that keeping the shop above 65°F in the spring and fall practically eliminates the problem.

High humidity in a basement shop occurs at a constant temperature, so condensation isn't an issue. The real problem here is that long-term high humidity will cause your wood

to absorb moisture. The joints that fit perfectly in your shop will open up when brought upstairs to drier conditions. Get a humidity gauge (see "Monitor the Moisture Levels in Your Wood" below) and monitor the levels in your shop from season to season. Daily fluctuations are not the problem—it's the trend. If humidity levels are consistently above 50%, run a dehumidifier.

A shop with low humidity is also a problem, unless you live in the desert. Low humidity levels are common in the winter in basement shops in cold climates—the low outside humidity combined with the nearby furnace can suck all the moisture out of your lumber. When your projects go where the ambient humidity is greater, the wood will absorb some moisture from the air and swell, blowing apart those finely fitted joints. If the humidity level in your shop is consistently below 25%, get a humidifier.

Heating

Adjusting moisture to take care of your tools and materials will keep you comfortable, except in winter. If you want to do any serious work in a garage shop in the winter,

MONITOR THE MOISTURE LEVELS IN YOUR WOOD

Knowing the humidity level is adequate for estimating the changes you can expect kiln-dried wood to go through when it leaves your shop. If you want to be scientific about it, or if you're air-drying lumber cut from your property, get a moisture meter. It measures the percentage of water in wood by evaluating its electrical resistance. The less expensive type of meter uses two metal prongs to test; the more expensive variety shown here does not damage the wood.

SHOP HEATERS

If you work in a garage shop in winter, heating is a very important issue. This chart shows the pros and cons of the most common types of heaters.

	PRO	CON
Propane job site heater	High Btus Quickly raises temperature Portable Not expensive	Adds some moisture to the air Might use a 20-gal. tank a weekend in cold climates Explosion hazard with gas Requires ventilation
Kerosene	Inexpensive Fuel-efficient	Odor and soot Requires ventilation Open flame Slow to raise temperature
Woodstove	Uses scraps Dry heat Emotionally satisfying	Heats slowly Requires tending Requires flue

Radiant propane heater	High Btus
	Can run with thermostat
	   for constant heating
	Large tank outside filled on
	   schedule by pros
		Pilot light explosion hazard
		   with solvent vapors

Electric	No open flame
	Even heat
	Can run with thermostat for
	   constant heating
	Built-in fan circulates air
		The most efficient heaters require
		   a 220v circuit

Shop as part of household heating system (with separate thermostat)	Probably most economical in the long run Constant temperature No open flame	Cost of installation

you'll need a heater. The chart above shows the pros and cons of various types of heaters. After years of using woodstoves and propane heaters and having to plan ahead to get the shop warm enough to work, I installed a 220v electric heater to keep the shop at 55°F. When I enter the shop in the morning, I turn up the heat a little, and it's soon warm enough to work in a sweater. It's a safe way to heat and no more expensive than propane or kerosene. It's a pleasure to walk into a warm-enough shop anytime I want to work. My only regret is that I didn't do it sooner.

The Well-Rounded Shop

What to Consider

Your shop is filled with equipment, and your skills have grown along with your tool inventory. At this point, there's not much you can't do. But a few more items will make your woodworking life more efficient and enjoyable.

The tools in this section are items you don't absolutely need to accomplish most woodworking tasks. They're nice to have, but they could be considered luxuries since you can do their functions with tools you already own. For instance, a set of dado blades for your tablesaw makes dadoes and grooves faster and more accurately than your router or hand tools can. Similarly, a dedicated mortising machine can turn you into a mortising fiend, but you could cut mortises by hand, with a router, or with an attachment on your drill press—not as efficiently, but you could cut them.

A portable dust collector moved from tool to tool will make a big difference in your quality of life. You'll spend less time cleaning up in the shop, and since less dust will make its way into your living area, you'll save on housework, too. Your planer will run more efficiently when hooked up to the collector, as shavings will not clog the cutterhead, and if you do much tablesaw work, you'll be relieved to be free of all that dust in your face.

While many schools of woodworking rely heavily on sanders, I don't. That's not to say sanders are bad or that you shouldn't own one until you're an accomplished woodworker. Oscillating-spindle and other sanders are in this section of the book simply because using a sander for the wrong task can ruin fine work. If you're early in your woodworking career and building a project that would benefit from an oscillating-spindle sander, you should have one. Saving these tools until last assumes you've gained the experience to understand the difference between a shortcut and a wrong turn.

A dedicated mortising machine does one thing: cuts square holes. It speeds up the task, but you can do the same thing with a router jig, with a drill press , or by hand.

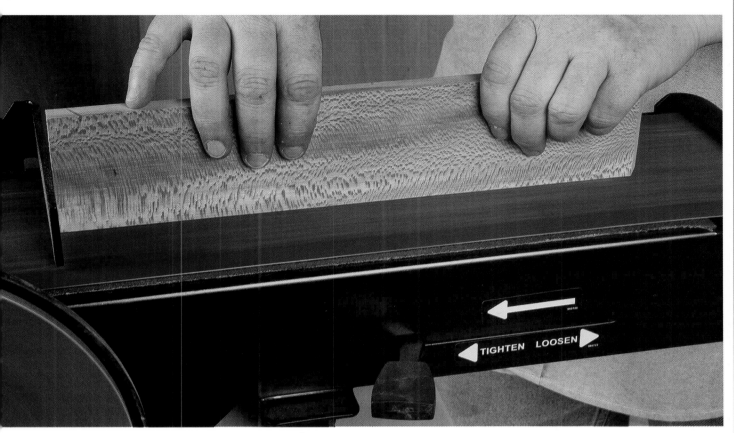

▲ **A belt/disk sander** is a powerful tool that can save time, but if you're not careful, it can quickly make hash of your work.

▶ **A dust collector** will add to the quality of your life. You'll spend less time cleaning up, and the air will be cleaner in both your shop and home.

Though the tablesaw appears in the previous section, more accessories to help you get the most from it appear in this one. They will certainly improve your tablesaw work, but they are not entry-level requirements.

The biscuit joiner is another tool that might have appeared in an earlier section. It's often considered a simple way to join wood, but getting consistently good results with it is difficult. That's not because the tool is ornery, but because it requires focus and organization to cut each piece with the correct orientation and alignment. Success with a biscuit joiner requires you think like a woodworker, and that takes some experience. By now you have it.

Dust Collector

Good dust management requires a three-tiered approach. Your shop vac is the first line of defense. Use it for general cleanup and attach it to sanders and circular saws, as well as the miter saw, the bandsaw, and the router table.

Planers, tablesaws, sanding machines, and other big dust producers require a high-volume dust collector designed for woodworking machinery. The third line is a ceiling-mounted air cleaner to scrub out the tiny particles (down to 1 micron) that slip through the collector bags.

What to buy

You can exhaustively analyze your dust-collection needs, but in a small shop, there's no need to make it complicated. If you use a mobile collector and hook it up to only one tool at a time using no more than 10' of hose, get a 1½-hp single-stage collector (see the photo on the facing page). If your space, budget, and wiring allow, go for the 220v, 2-hp model. Most of the dust drops into the lower bag, while the permeable upper bag filters finer particles as the air blows through it. The standard upper bag only collects particles larger than 30 microns. Be sure to buy the optional 5-micron filter bag or one of the efficient new filtration canisters.

A typical home shop about the size of a two-car garage needs an air cleaner capable of moving 1,000 cubic feet of air per minute (cfm) (see the photo on the facing page). Make sure it has slower speeds as well—1,000 cfm is noisy. Opt for the washable electrostatic prefilter, and hold out for a remote control to avoid reaching for the switch constantly. Another great feature is a timer to turn off the cleaner after a couple of hours so you don't have to remember to do so.

WHAT DUST COLLECTORS CAN DO

■ COLLECT CHIPS AND DUST

Use large-diameter hose right up to the machine.

■ CONNECTS

Use large ports to connect directly to machines.

■ SEPARATES

A trash can with a special lid keeps big chips and debris from damaging the fan blade.

Single-Stage Dust Collector and Air Cleaner

For most small shops, a 1½-hp single-stage dust collector is all you need. Move it from tool to tool as needed.

An air cleaner will keep the shop noticeably cleaner. If your shop is connected to your house, it'll keep the house cleaner as well—less dust on the piano. If possible, hang it overhead a few feet from a side wall to promote air circulation.

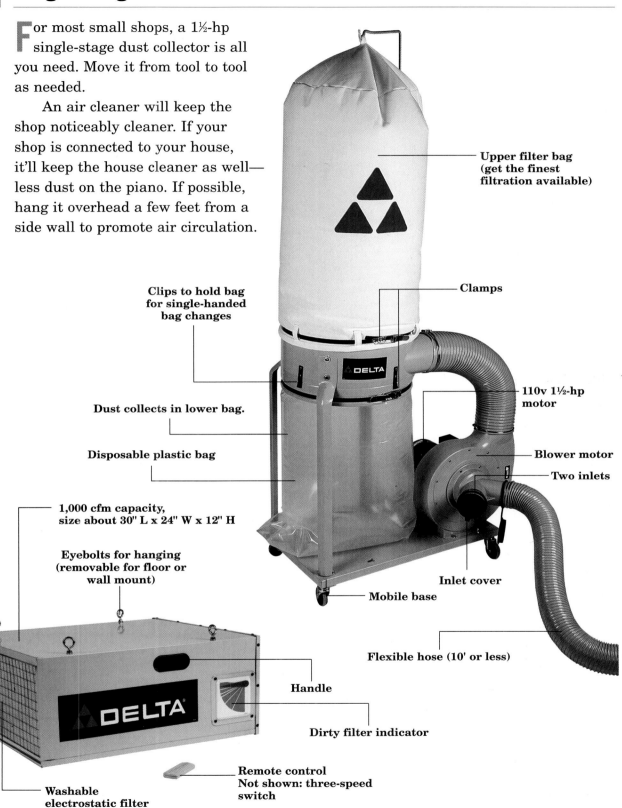

Upper filter bag (get the finest filtration available)

Clips to hold bag for single-handed bag changes

Clamps

Dust collects in lower bag.

Disposable plastic bag

110v 1½-hp motor

Blower motor

Two inlets

1,000 cfm capacity, size about 30" L x 24" W x 12" H

Inlet cover

Mobile base

Eyebolts for hanging (removable for floor or wall mount)

Flexible hose (10' or less)

Handle

Dirty filter indicator

Washable electrostatic filter

Remote control Not shown: three-speed switch

Sanders

When you mention sanders, most people think of disk or random-orbit sanders, but there are a number of other sanding tools designed to tackle specialized jobs more efficiently. Admittedly, these sanders are not vital tools—it makes sense to buy them on an as-needed basis. But once you own any one of them, you'll find it solves all kinds of problems you didn't know you had.

DRUM THICKNESS SANDER

Like a planer, a thickness sander removes material from the top of the board. But rather than cutting with knives, sandpaper wound around the drum abrades the surface of the board as it moves on a conveyor belt.

A thickness sander with a cantilevered drum like the one shown here on p. 140 has a sanding capacity twice the length of its drum. You simply run one half of the workpiece under the drum, then flip it around and run the other.

The thickness sander aces some of the planer's biggest problems. First, there's no tearout—no matter how wildly figured the grain. Second, the sander is more accurate than a planer and is ideal for dimensioning stock to less than ¼". Third, the conveyor belt makes it safe to run small pieces through the machine.

What to buy

Choose a sander with a cantilevered head for maximum capacity. Look for a large dust-collection port and casters or a mobile base. Securing the sandpaper to the drum is the most difficult part of using these machines; avoid any that have fussy little clips tucked under the end of the drum. Remember that a cantilevered head is prone to sagging out of parallel to the table, and you'll have to check and adjust it often when sanding wide panels. The simpler the adjustment process, the better.

WHAT SANDERS CAN DO

■ THICKNESS WIDE BOARDS

Handle boards up to twice the length of the drum.

■ SAND INSIDE CURVES

Smooth and fair inside curves.

■ SAND OUTSIDE CURVES

Smooth and fair outside curves.

The Detail Sander

With its compact triangular head and aggressive sanding pattern, a detail sander can smooth and profile wood in places no other tool can reach. This model accepts scraper and saw blades for even greater versatility.

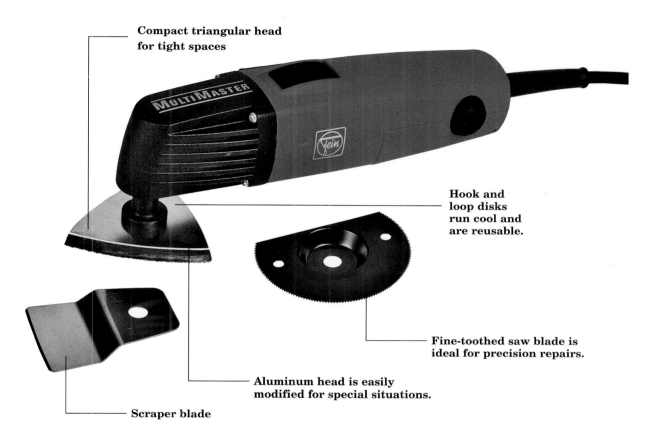

Compact triangular head for tight spaces

Hook and loop disks run cool and are reusable.

Fine-toothed saw blade is ideal for precision repairs.

Aluminum head is easily modified for special situations.

Scraper blade

■ **CREATE ROUNDOVERS**

Turn sharp edges into rounded corners.

■ **SAND BEVELS**

Tilt the table to sand a bevel with disk or belt.

■ **REMOVE METAL**

Use the belt or disk sander to modify metal tools and parts.

OSCILLATING-SPINDLE SANDER

The oscillating-spindle sander excels at sanding curves. A rubber drum with an abrasive sleeve mounts in the center of a table, and it both rotates and moves up and down when running. The movement in two directions means that the machine runs cooler, distributes the wear over a greater area, and produces a smoother surface by randomizing the cutting pattern (see the bottom photo on the facing page). To get a smooth curve on a worksurface, use a sanding drum with a diameter that suits the radius you're sanding.

What to buy

Get a sander with the longest stroke because it will distribute wear better. Look for a good dust-collection port, and even if it comes with a cloth dust bag, plan on using a shop vacuum with it. A tilting table is not crucial but will increase the versatility of your sander, which is always a good thing.

BELT/DISK SANDER

This versatile sander solves many little problems around the shop—rounding corners, flattening an edge, or smoothing the face of a

small workpiece. It's also handy for shaping and smoothing metal pieces.

The belt and disk share both the motor and the table. The belt can be horizontal, where it's best suited for flattening and smoothing. The table in front of the disk makes it easy to round corners, and if you use the miter gauge in the slot, you can easily square ends. The table also angles for bevels. Or, you can switch the belt to the vertical position and install the table in front of it for working on longer pieces.

WHAT SANDERS CAN DO (continued)

■ FLATTEN AND SMOOTH

The base plate of the belt sander flattens and smooths small pieces.

■ SAND IN CLOSE QUARTERS

A detail sander can work places no other sander can reach.

A Pair of Useful Sanders

The benchtop oscillating-spindle sander on the left is best suited for handling curves, while the 6" belt/9" disk sander on the right is good for rounding corners and straightening and smoothing edges.

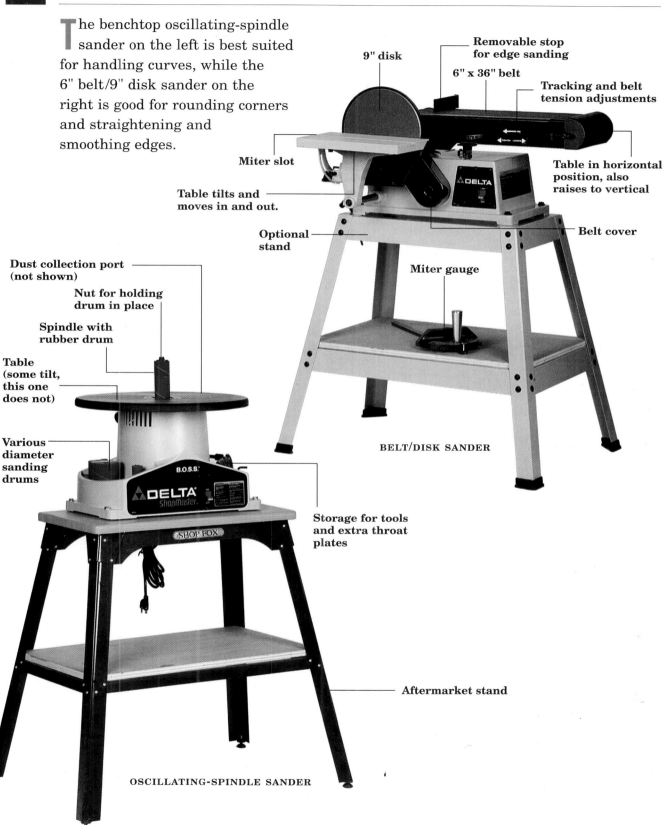

9" disk

Removable stop for edge sanding

6" x 36" belt

Tracking and belt tension adjustments

Miter slot

Table tilts and moves in and out.

Table in horizontal position, also raises to vertical

Optional stand

Belt cover

Miter gauge

BELT/DISK SANDER

Dust collection port (not shown)

Nut for holding drum in place

Spindle with rubber drum

Table (some tilt, this one does not)

Various diameter sanding drums

Storage for tools and extra throat plates

Aftermarket stand

OSCILLATING-SPINDLE SANDER

The Cantilever Drum Thickness Sander

Accurate, safe, and virtually tearout free, this thickness sander can solve many problems around the shop and is a good complement to a thickness planer.

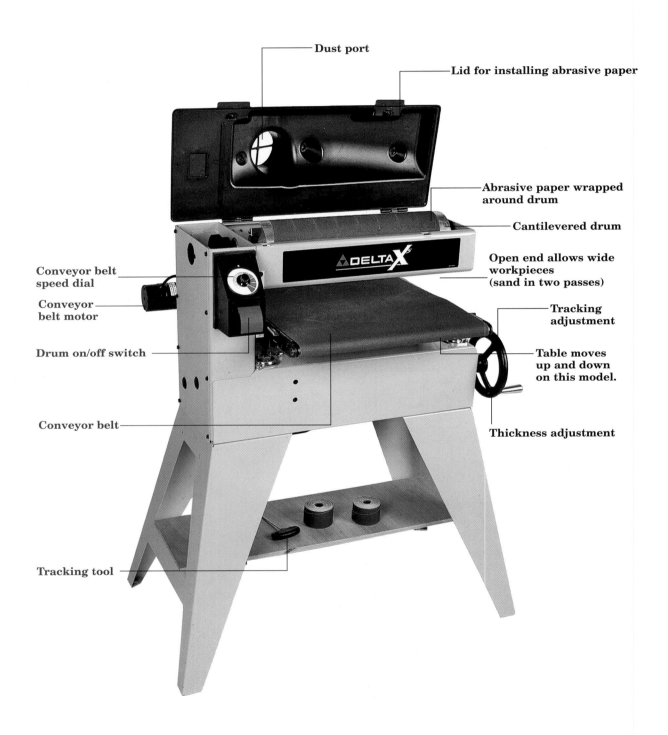

Dust port

Lid for installing abrasive paper

Abrasive paper wrapped around drum

Cantilevered drum

Conveyor belt speed dial

Conveyor belt motor

Drum on/off switch

Conveyor belt

Open end allows wide workpieces (sand in two passes)

Tracking adjustment

Table moves up and down on this model.

Thickness adjustment

Tracking tool

Even a small belt sander is a brute of a tool. It wants to run away from you, and all but the strongest and most experienced users spend lots of energy just trying to keep it in check—forget any kind of fine-tuning to make sure the resulting surface is flat and square.

If you're going to use a belt sander, know that they are notoriously unbalanced tools. Make sure you're working at a convenient height with work clamped in place, and position yourself where it's easy to keep the sander horizontal. Don't linger in any one spot for too long, and maintain your position as you move along the surface.

There's no question that the handheld belt sander is a powerful tool, one that can do fine work. The problem is that people tend to rely on it for more than can be expected of either the tool or the user.

What to buy

A 6" belt/9" disk sander is a good size for most small shops—a good thing, because the price jumps considerably for the next size up. Look for a sander with a large table and easy-to-operate adjustments (see the top photo on p. 139). Make sure the belt-tracking mechanism is easy to reach in both vertical and horizontal positions—it's the adjustment you'll make most.

DETAIL SANDER

A nice little triangle sander with hook-and-loop sandpaper is the best tool for refinishing and sanding in difficult locations. And there are some situations, such as when you are making louvered doors, when nothing else will do the job.

Most of these sanders vibrate at around 20,000 rpm, and they can leave a persistent and distinctive scratch pattern. You'll have to keep a full array of grits on hand and progress your way through them to end up with a truly smooth surface.

What to buy

Look for a heavy-duty machine; a low-cost model may not last through one tough job. Pick a brand that offers a variety of pad shapes and make sure they're easy to get locally. Aluminum pads are easy to modify to suit a job. You can fit some brands with a saw or scraper in lieu of the sanding pad (see the top photo on p. 137), invaluable for surgical repairs or demolition. Finally, get a brand that has dust collection or a dust-collection kit you can add on later.

Biscuit Joiner

Using a biscuit joiner is a quick way to make strong, good-looking joints. The tool is essentially a horizontal saw that cuts matching arcs into the mating edges of two boards. The biscuit is a football-shaped piece of pressed wood that slips into the slots in each board and bridges the structural gap between them. Properly glued and clamped, the resulting joints are very strong, but getting that result requires a little skill and attention. Biscuit joiners are best used for reinforcing glued joints that aren't inherently strong, such as a butted corner joint or just about any joint in plywood (see the table on p. 144).

The most common error in cutting biscuit joints is losing track of the reference surface. Once the slot height is set, all cuts must be made with the fence on the same reference surface—usually the outside face of the workpiece. If you mess up along the way and cut a

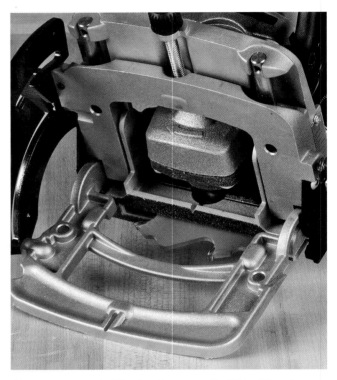

Cutter exposed. The business end of the biscuit joiner shows the horizontal blade in the cutting position.

WHAT A BISCUIT JOINER CAN DO

■ DOUBLE-LOOSE TENON

Multiple biscuits are easily installed and make one of the strongest joints in woodworking.

■ BUTTED CORNER

A simple and secure way to join a box or drawer.

■ FACE-FRAME JOINT

Join two pieces in a T joint anywhere along their length.

Biscuit Joiner

A biscuit joiner is a good way to reinforce joinery, especially when building in plywood.

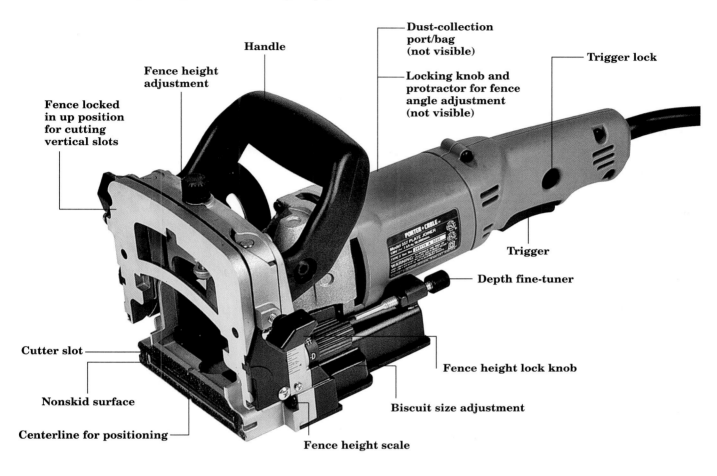

Handle

Fence height adjustment

Fence locked in up position for cutting vertical slots

Dust-collection port/bag (not visible)

Locking knob and protractor for fence angle adjustment (not visible)

Trigger lock

Trigger

Depth fine-tuner

Cutter slot

Nonskid surface

Centerline for positioning

Fence height scale

Biscuit size adjustment

Fence height lock knob

■ BOOKSHELF JOINT

A butted joint doesn't have to be at the corner.

■ ANGLED JOINT

With the right fence, you can cut slots at any angle.

■ EDGE TRIM

Cover unsightly plywood edges by biscuiting solid wood in place.

BISCUIT JOINER VS. POCKET-HOLE SCREWS		
	BISCUIT JOINER	**POCKET-HOLE SCREWS**
PROS	Makes invisible joints Is easier for large workpieces and sheet goods	Are quick Holes can be made on site with mini-jig Are removable Don't require clamps while glue dries
CONS	Requires clamping while glue cures Requires care with reference surfaces or alignment can be a problem	Are visible Can be difficult to position large pieces in jig

few with the fence against the inner surface, slots won't match up. Another alignment problem results when the fence and base don't solidly contact the workpiece during the cut. The slot won't be parallel to the surface, and the angled biscuit won't slip into the mating slot. Prevent this with proper stance and by clamping the workpiece so it can't move during the cut (see the top photo on the facing page).

Such errors are so common it's surprising that the biscuit joiner is often hailed as an alignment tool, just the thing for registering boards when gluing up a panel. It's not. Edge-

to-edge joints made with modern glue are so strong that biscuits are overkill: There's no need to spend time fussing over their alignment.

What to buy

For the most versatile tool, look for one that can handle a range of biscuit sizes. The smallest #FF face-frame biscuits are handy when you're working with 1½"-wide material, and not all machines can cut such small slots. Some machines have settings to cut Simplex and Duplex slots, an industry standard used for hinges and knockdown hardware. Next on

WHAT A BISCUIT JOINER CAN DO (continued)

■ MOLDING AND TRIM

Biscuit joining molding and trim keeps corners jointed tight.

■ MORTISE FOR HARDWARE

A Z-shaped tabletop fastener is screwed to a tabletop and slipped into a slot on the apron.

Proper approach is key. Alignment is everything in biscuit joining. Proper stance keeps the tool horizontal and in firm contact with the reference surface (in this case, the top of the workpiece).

your list should be the quality of the fence and its adjustments. Don't settle for anything less than rack-and-pinion adjustment, and make sure you can lock the fence in place so it won't move in use. Look at the height adjustment scale and make sure it's clear and easy to read. Finally, check to see that the handle and on/off switch are comfortable and convenient in your hands.

Choose the right biscuit. Made of compressed wood dust, biscuits lock the joint when the moisture in glue causes them to swell. From top to bottom: #10, #0, and #FF (face frame).

Tablesaw Accessories

A s you get more familiar with your tablesaw, you'll find ways to improve its safety, versatility, and performance. The list is endless, but here are a few accessories that will give you the biggest bang for your buck.

Safety

The simple guard that comes on most saws leaves something to be desired. An overarm guard is much better—it doesn't interfere with everyday operations and is easily removed when it does. The separate splitter virtually eliminates the risk of kickback, and a dust-collection port on the guard collects dust before it falls into the saw. Collect the rest of the dust beneath the saw in a bag, or get a brand-specific plastic shield with a dust port that seals the underside of the saw. For maximum dust collection, don't forget to seal the back of the saw around the motor with a plywood cutout taped or screwed in place (just visible in the bottom center photo on p. 148).

Performance

Help your saw work harder and smarter by replacing the V-belt with a link belt. Not only will it increase the power transfer from motor to blade, but it will also reduce vibration. An outfeed table helps that last bit of follow-through when you're feeding stock and makes your work more secure—even a small shop-made table is a big improvement.

A zero-clearance insert is blank when you install it. Retract the blade, clamp the insert in place, and slowly raise the blade. You'll wind up with a perfectly fitting slot that not only reduces tearout on the underside of a board, but also increases dust-collection efficiency. Since the inserts are not interchangeable,

WHAT THESE ACCESSORIES CAN DO

■ **AFTERMARKET MITER GAUGE**

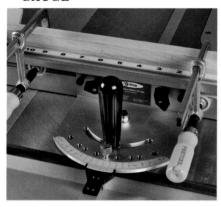

A quality miter gauge has a sliding fence with vertical faces for secure clamping.

■ **SPLITTER AND ZERO-CLEARANCE THROAT PLATE**

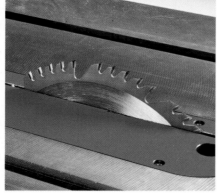

A zero-clearance throat plate reduces tearout, and a splitter virtually eliminates kickback.

■ **OVERARM BLADE GUARD WITH DUST COLLECTION**

This accessory protects your hands and collects the lion's share of the sawdust.

BLADES FOR YOUR TABLESAW

The single most important upgrade you can make to your tablesaw is to install a high-quality blade. The increased performance is stunning. You'll get less noise and vibration and a smoother, faster cut.

You could go into an exhaustive study of the geometry and physics of tablesaw blades, but within a price range, the resulting cuts are remarkably similar, even if the blades look different. You can't go wrong with an upper-middle-price-range blade from a well-known manufacturer.

A **combination blade** will handle almost all of your tablesaw work. Get one with 30 or 40 carbide-tipped teeth. Get two if you can, so you can continue working while one blade is being sharpened.

If you're ripping a significant quantity of solid wood, use a carbide-tipped **rip blade** with about 20 tpi. With fewer teeth, it'll cut faster and cleaner.

A **dado set** allows you to cut wide grooves in one pass (see the bottom right photo on p. 148). It consists of a set of matched blades and shims designed to be stacked in various combinations to equal a desired width. Dado blades are smaller in diameter than standard tablesaw blades, with less cutting height. Get an 8" set for versatility.

A **blade stabilizer** keeps a thin kerf-blade running truer by reducing vibration. It also decreases the usable blade height.

start out with one for your most frequently used blade, adding more as you see fit. You can also make one to use when cutting 45° (or any other) angles.

A tablesaw is as accurate as the jigs used on it, and you can't beat a cast-iron tenoning jig for sheer mass or adjustability. It slides in a miter slot, holds workpieces vertically, and uses fine-thread screws to adjust the workpiece's distance from the cutter. For cutting tapered legs fast, use a tapering jig. The best tapering jigs ride in the miter slot and hold the workpiece at an angle as it passes through the blade.

■ TENONING JIG

Heavy, secure, and highly adjustable, this jig makes cutting tenons and other vertical cuts a breeze.

■ TAPERING JIG

This jig rides in the miter slot and holds the workpiece securely at an angle.

■ OUTFEED EXTENSION

Even small shopmade outfeed tables provide a lot more support at the end of a cut.

Besides accuracy, its greatest benefit is that the work doesn't slide on the saw's table. Instead, workpieces are supported on the sled, which means that there's no chance of them twisting and causing a kickback. Using a shopmade sled is the safest and most accurate way to crosscut.

You can make a simple crosscut sled that rides in one miter slot, but you'll get more use out of a moderately sized one that rides in both miter slots. The key to accuracy in each case is that the fence on the user's side of the blade is perpendicular to the slots (see the top phot at right).

In time, you'll have a collection of sleds—from small job-specific, single runner sleds to massive boxes capable of handling 24"-wide sheets of plywood, plus sleds for beveled cuts, dadoes, and mitering (see the bottom photo at right).

Work safely. A crosscut sled is safer and more accurate than a miter gauge because the wood doesn't slide. Hold the sled with two hands well away from the blade.

Ease large jobs. Use a crosscut sled when working with wide or heavy pieces. Hold them down and firmly against the fence.

WHAT THESE ACCESSORIES CAN DO (continued)

■ LINK BELT

Replace the V-belt with a link belt and increase the drive power.

■ PAL ALIGNERS

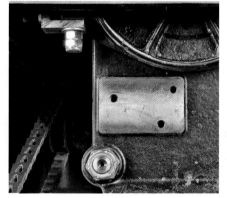

Setscrews threaded into two aluminum els push against motor mounting bolts to help align a contractor's saw.

■ SET OF DADO BLADES

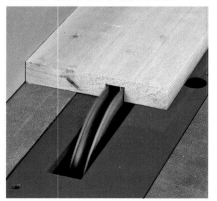

A dado set contains blades and shims you can gang up to cut grooves of a specified width.

Tablesaw Upgrades

A tricked-out tablesaw is safer, more accurate, and more versatile. This is only the beginning of what you can do to get more use out of your tablesaw.

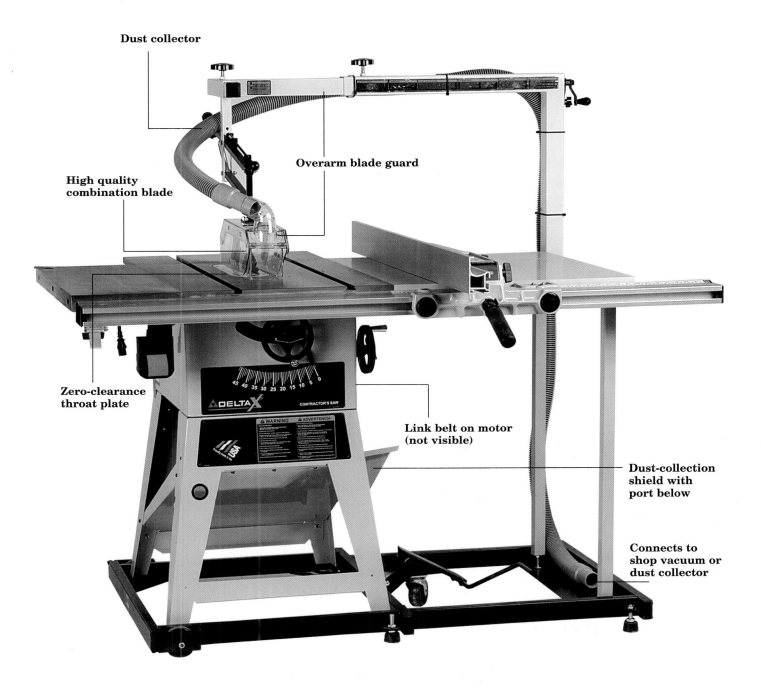

Dust collector

Overarm blade guard

High quality combination blade

Zero-clearance throat plate

Link belt on motor (not visible)

Dust-collection shield with port below

Connects to shop vacuum or dust collector

Hollow-Chisel Mortiser

You can add a mortising attachment to your drill press, but if you're going to cut more than a few mortises, you need a mortising machine. It's nothing more than a drill press optimized for drilling a series of square holes. A hollow-chisel mortiser offers a limited depth of cut, a powerful lever handle, and an adjustable fence and hold-down for positioning the work. To cut square holes, you'll also need an interesting bit called a hollow chisel.

The outer (hollow) portion is square in section with a chisel-sharp perimeter at the bottom. An aggressive drill bit runs inside this hollow chisel; its outside diameter is very close to the inside diameter of the hollow portion. The drill bit removes the bulk of the waste (it's ejected through a slot in the side of the hollow chisel) while the hollow chisel removes only a little at the corners to square up the hole.

Cutting a mortise. A hollow-chisel mortiser cuts square holes with ease. Line up a series of holes to make a mortise.

WHAT A HOLLOW-CHISEL MORTISER CAN DO

■ CUT SQUARE MORTISES

Hollow chisels cut mortises to accept tenons for joinery.

■ CUT THROUGH MORTISES

For a decorative look—and more strength—cut joints where tenons go all the way through.

Benchtop Hollow-Chisel Mortiser and Chisels

A hollow-chisel mortiser is nothing more than a drill press made for cutting square holes.

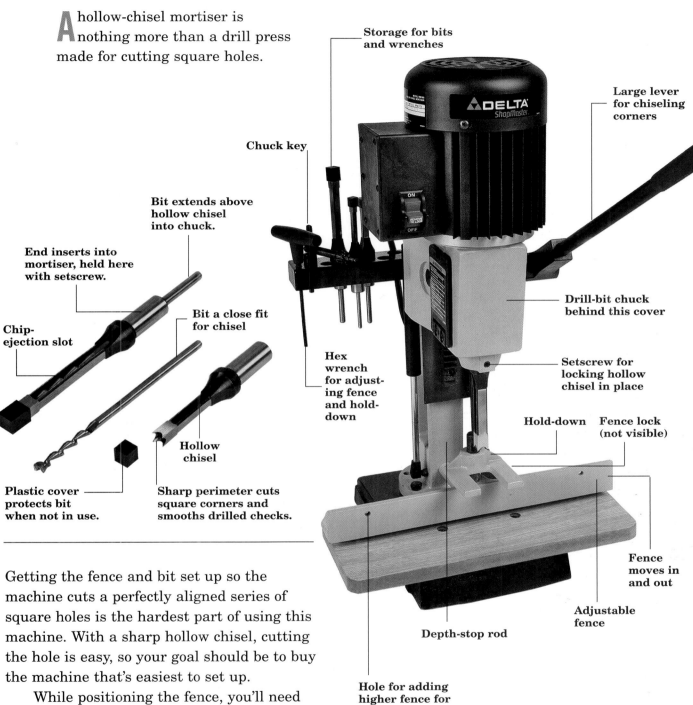

Storage for bits and wrenches

Large lever for chiseling corners

Chuck key

Bit extends above hollow chisel into chuck.

End inserts into mortiser, held here with setscrew.

Chip-ejection slot

Bit a close fit for chisel

Drill-bit chuck behind this cover

Hex wrench for adjusting fence and hold-down

Setscrew for locking hollow chisel in place

Hollow chisel

Hold-down

Fence lock (not visible)

Plastic cover protects bit when not in use.

Sharp perimeter cuts square corners and smooths drilled checks.

Fence moves in and out

Adjustable fence

Depth-stop rod

Hole for adding higher fence for larger workpieces

Getting the fence and bit set up so the machine cuts a perfectly aligned series of square holes is the hardest part of using this machine. With a sharp hollow chisel, cutting the hole is easy, so your goal should be to buy the machine that's easiest to set up.

While positioning the fence, you'll need to move it in and out in small increments. Look for an easy way to tighten and loosen the lock nut for quick adjustments. You'll also have to get the holddown tight but not too tight—so look for easy lock/unlock there, too.

The Well-Rounded Shop Space

After working with the tools introduced in the Efficient Shop, you'll have some strong ideas about how to improve your setup. As you add the specialized tools in this section—the Well-Rounded Shop— you'll probably reach a turning point at which it becomes necessary for you to take some time just to work on your shop (see the photo on the facing page).

You'll find you need to think about the allotment of horizontal surfaces in your shop. You've probably started doing certain tasks in the same part of the shop every time—take note of what wants to happen where and set up areas to help that work go more smoothly. Build racks or drawers to hold the tools you'll need, upgrade the lighting, and organize the space so that what you need to do the job is always at hand.

You'll need a sharpening area, a place where having water and metal slurry slopped about won't be a problem. Put down a section of melamine or plywood and clear the area so overspray from your bottle of water won't cause rust.

Most shops end up with a drilling area near the drill press. Build racks and holders for your bits and set up a charging area for the cordless drills.

Set aside a corner of the shop as a metalworking area. Get an inexpensive metalworking vise (weight counts for everything here) and bolt it to the bench. Use it to hold metal you need to manipulate for whatever reason— threaded rod or bolts you're cutting, metal

Every woodworking shop needs a metalworking vise for gripping items that would damage the wooden cheeks of a woodworking vise. Bolt it to the counter in an area where rusty filings won't reach fine work.

plates for outdoor construction, lawn-mower parts, and the like.

A couple of sets of specialized sawhorses and some plywood supply horizontal space on demand and are out of the way otherwise. Build a set as high as your tablesaw, and you can have an infeed or outfeed table (or both) large enough to support a full sheet of plywood. A low set, about 18" high, serves as a bench at a convenient height for assembling and finishing projects too large or high to put on your workbench.

Floor Plan, the Well-Rounded Shop

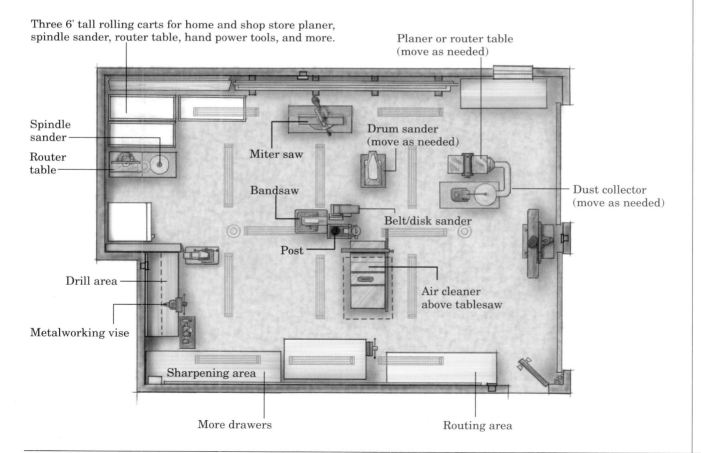

Three 6' tall rolling carts for home and shop store planer, spindle sander, router table, hand power tools, and more.

Planer or router table
(move as needed)

Spindle sander

Router table

Miter saw

Drum sander
(move as needed)

Bandsaw

Dust collector
(move as needed)

Belt/disk sander

Post

Drill area

Metalworking vise

Air cleaner
above tablesaw

Sharpening area

More drawers

Routing area

Periodically in your life as a woodworker, you'll find you must work on your shop. Your increasing sophistication places increasing demands on the shop, and with your increased skills you can make it right. When you're done with the upgrade, you're ready to work at a new level.

Though no small shop can ever manage a truly dust-free finishing room, you can still set up one area of your shop for finishing. Keep it near a window or door so you can set up a fan to vent the solvent fumes, and try to keep the area around it free of clutter so it's easy to vacuum clean. Plastic sheeting hung from a framework built into the overhead helps keep dust out of the rest of the shop.

With so many machines in your shop, it may be time to think of an annex. A shed can take care of the garden tools, some of the finishing materials, and perhaps some household items. A lean-to beneath the eaves is a great place to store lumber. And just outside the garage door you could build a good-looking locker to store infrequently used tools.

With your increased skills and shop capacity, you can design and build storage to fit in any space. Build a chest of drawers to fit under the right side extension of your tablesaw. Replace some of your tool stands with a cabinet of drawers on casters. Or build a bank of drawers to handle all the small stuff.

We need our shops to do woodworking, and so we go to work on our shops. We add storage, move a tool, or build something unique, and soon working on the shop takes on a life of its own. All woodworkers go through periods where their major project is the shop. They come out of it ready to tackle new levels of woodworking. Skill and tools—when you improve one, you improve the other.

Index